U0174681

2019北京国际首饰艺术展

2019 BEIJING INTERNATIONAL JEWELLERY ART EXHIBITION

主　编 | 詹炳宏
EDITOR IN CHIEF | ZHAN BINGHONG

副主编 | 兰翠芹 | 高伟 | 胡俊
DEPUTY EDITOR IN CHIEF | LAN CUIQIN | GAO WEI | HU JUN

中国纺织出版社有限公司

内 容 提 要

2019北京国际首饰艺术展以“贯通”为主题，旨在展现不同文化的交流与碰撞，推进多元化首饰艺术创作潮流，并为不同的艺术创作观念和思潮搭建互动的平台与空间。本次展览共收到全球46个国家和地区658位艺术家的作品，最终评选出312位作者参展。本书精心挑选其中具有代表性的艺术作品，力求展现展览的引领性、多样性以及启发性。

本书图文并茂，图例丰富，适合高等院校珠宝首饰专业师生、珠宝首饰设计师、收藏家以及广大首饰爱好者阅读与参考。

图书在版编目（CIP）数据

2019北京国际首饰艺术展：汉英对照/詹炳宏主编. -- 北京：中国纺织出版社有限公司，2021.9

ISBN 978-7-5180-8743-3

Ⅰ. ①2… Ⅱ. ①詹… Ⅲ. ①首饰－设计－作品集－世界－现代 Ⅳ. ①TS934.3

中国版本图书馆CIP数据核字（2021）第150988号

责任编辑：李春奕　　特约编辑：徐铭爽
责任校对：楼旭红　　责任印制：王艳丽

中国纺织出版社有限公司出版发行
地址：北京市朝阳区百子湾东里A407号楼　邮政编码：100124
销售电话：010—67004422　传真：010—87155801
http://www.c-textilep.com
中国纺织出版社天猫旗舰店
官方微博 http://weibo.com/2119887771
北京华联印刷有限公司印刷　各地新华书店经销
2021年9月第1版第1次印刷
开本：787×1192　1/8　印张：31
字数：228千字　定价：468.00元

INTEGRATION & CONNECTION

北 京 国 际 首 饰 艺 术 展 · **贯 通**

2019 BEIJING INTERNATIONAL
JEWELLERY ART EXHIBITION

2019 北京国际首饰艺术作品展

展览汇聚了来自全球 46 个国家和地区 312 位艺术家、设计师的作品。其中包括英国伯明翰艺术与设计学院珠宝学院院长 Stephen Bottomley、美国俄亥俄州博林格林州立大学教授 Tom Muir、意大利当代首饰艺术学会主席 Maria Rosa Franzin 等国际著名首饰艺术家。

北京国际首饰艺术展开幕式及参展作品秀

区别于国际首饰艺术静态展，作品秀以动态时尚走秀的形式呈现世界各国艺术家及设计师的首饰作品。

第四届北京国际首饰艺术展高峰论坛

该论坛主要聚焦探讨首饰艺术的可能性，研究首饰艺术与外部世界、数字化与人工智能等新兴科技间的联系与互动。

国际首饰设计高校联盟成立大会

来自全球的将近 40 所首饰设计院校加入了国际首饰设计高校联盟的大家庭，为国际首饰设计教育贡献力量。

国际首饰设计教育座谈会

针对首饰艺术教育话题，探讨国际语境下的首饰艺术以及商业首饰的发展方向、趋势和未来首饰设计人才发展需求。

2019 北京国际首饰艺术展旨在展现世界首饰艺术与设计前沿理论和产业发展趋势，艺术展汇聚了全球最具影响力的学者、专家、企业家，是在中国举办的国际最高水平首饰艺术与设计的学术活动。北京国际首饰艺术展为学者、设计师们搭建了引领国际首饰设计前沿学术思想、理论和实践研究成果的交流、发布平台，艺术家们可在此共同研究、探索在当今经济全球化、市场共享化的新时代，首饰艺术与设计如何连接政治、经济、人文，并与相关产业融合发展、实现科技成果转化，为促进世界文化发展贡献力量。

INTEGRATION & CONNECTION

本次展览的主题定为“贯 通”，其中“贯”，意为“穿，通，连”之意；“通”，本意为通达、没有障碍，后引申为到达目的地，又引申为互相连接无阻断。也指“使知道”“传达于对方”，或“了解”“懂得”的意思。引申为通晓首饰艺术的精妙之境，打破首饰艺术的壁垒，不同文化之间取长补短、相互促进，以及不断推陈出新的意思。

The theme of the Exhibit on is “贯 通” (“Integration and Connection”). In Chinese, the character “贯（guan）” means going through, being connected or joined together. The character “通（tong）” originally means clear and without obstacles. Later it was used to mean arrival at the destination, or extended further to mean being connected and without interruptions. “通（tong）” also means to let someone become aware of something, to pass the message over, or to comprehend. For the Exhibition, the theme is selected to illustrate the delicacy in understanding the art of jewellery and the lift of restrictions. It also accentuates the importance for different cultures to learn from each other, thrive together, and abandon the old and nurture the new.

组织方式 | ORGANIZATION MODE

指导单位：

教育部中外人文交流中心

工业和信息化部工业文化发展中心

北京国际设计周组委会

承办单位：

北京服装学院服饰艺术与工程学院

北京服装学院中国生活方式设计研究院

北京服装学院艺术设计学院

中关村时尚产业创新园

协办单位：

北京设计学会

上海市首饰设计协会

《设计》杂志社 《雕塑》杂志社

主办单位：

北京服装学院

支持单位：

外国院校：英国皇家艺术学院、英国伦敦艺术大学、英国布莱顿大学、英国曼彻斯特都市大学、英国伯明翰城市大学、意大利佛罗伦萨阿契米亚首饰学院、意大利佛罗伦萨欧纳菲首饰学院、意大利帕多瓦塞瓦蒂克国立艺术学院、比利时安特卫普皇家美术学院、德国慕尼黑美术学院、德国普福尔茨海姆应用技术大学、西班牙巴塞罗那玛萨纳艺术学院、美国罗德岛设计学院

中国院校：清华大学美术学院、中央美术学院、中国地质大学、北京工业大学艺术设计学院、天津美术学院、中国美术学院、上海大学美术学院、南京艺术学院、山东工艺美术学院

企业公司：中港一带一路国际文化中心、北京菜市口百货股份有限公司、北京珐琅厂有限责任公司、北京金一文化发展股份有限公司、北京玉尊源玉雕艺术有限责任公司、萃华金银珠宝股份有限公司、金银珠宝股份有限公司、国金黄金集团有限公司、上海老凤祥有限公司、上海豫园黄金珠宝集团有限公司、上海铂利德钻石有限公司、上海华泰珠宝商场有限公司、上海张铁军珠宝集团有限公司、南京通灵珠宝股份有限公司、周大福珠宝金行有限公司、周生生珠宝金行有限公司、周大生珠宝股份有限公司、广东潮宏基实业股份有限公司、浙江天使之泪珍珠股份有限公司、深圳市百泰珠宝首饰有限公司、深圳市粤豪珠宝有限公司、深圳市甘露珠宝首饰有限公司（爱得康）、深圳市沃尔弗斯实业有限公司

Steering organizations:
China Center for International People-to-people Exchange Industrial Culture Development Center of Ministry of Industry and Information Technology
Executive Committee of Beijing Design Week

Organizers:
Fashion Accessory Art and Engineering College,BIFT
Chinese Academy of Lifestyle Design,BIFT
School of Art and Design,BIFT
BIFT Park

Co-organizers:
Beijing Design Society
Shanghai Jewellery Design Association
Design Magazine Sculpture Magazine

Sponsor:
Beijing Institute of Fashion Technology

Supporting organizations:

Foreign colleges and universities: Royal College of Art (UK), University of the Arts London (UK), University of Brighton (UK), Manchester Metropolitan University (UK), Birmingham City University (UK), Alchimia Contemporary Jewellery School (Florence, Italy), Le Arti Orafe Jewellery School & Academy (Florence, Italy), Pietro Selvatico Art Institute (Padua, Italy), Royal Academy of Fine Arts (Antwerp, Belgium), Academy of Fine Arts (Munich), Pforzheim University of Applied Sciences (Germany), ESTUDIO CARME PINÓS (Spain) and Rhode Island School of Design (US)

China's colleges and universities: Academy of Arts & Design of Tsinghua University, Central Academy of Fine Arts, China University of Geosciences, College of Art and Design of Beijing University of Technology, Tianjin Academy of Fine Arts, China Academy of Art, Shanghai Academy of Fine Arts of Shanghai University, Nanjing University of the Arts, Shandong University of Art & Design

Enterprise companies: Belt and Road International Culture Center, Beijing Caishikou Department Store Co., Ltd., Beijing Enamel Factory Co., Ltd., Beijing Kingee Culture Development Co., Ltd., Beijing Yu Zun Yuan Jade Sculpture Co., Ltd., Cuihua Gold, Silver and Jewellery Co., Ltd., Gold, Silver and Jewellery Co., Ltd., Guojin Gold Group Co., Ltd., Shanghai Lao Feng Xiang Co., Ltd., Shanghai Yuyuan Gold and Jewellery Group Co., Ltd., Shanghai Bolide Diamond Co., Ltd., Shanghai Huatai Jewellery Department Store Co., Ltd., Shanghai Zhang Tie Jun Jewellery Group Co., Ltd., Nanjing Tesiro Jewellery Co., Ltd., Chow Tai Fook Jewellery Group Ltd., Chou Sang Sang Jewellery Co., Ltd., Chow Tai Seng Jewellery Co., Ltd., Guangdong CHJ Jewellery Co., Ltd., Zhejiang Angeperle Pearl Co., Ltd., Shenzhen Batar Investment Holding Group Co., Ltd., Shenzhen Yuehao Jewellery Co., Ltd., Shenzhen Ganlu Jewellery Co., Ltd., Shenzhen Wolfers Industry Co., Ltd.

组　织 委员会 | ORGANIZING COMMITTEE

主任：

马胜杰（北京服装学院校务委员会主席）

罗民（工业和信息化部工业文化发展中心主任）

副秘书长：

马凤宝、兰翠芹、常炜

委员：

Ann Priest（英国）、白静宜、包晓莹、储卫民、程学林、陈国珍、陈晓华、Eimera Conyard(爱尔兰)、郭强、郭英杰、郭新、高伟、洪兴宇、Hans Stofer(英国)、胡书刚、黄晓望、黄雯、Lauret-Max De Cock（比利时）、Leo Caballero（西班牙）、李英杰、李雪梅、廖创宾、马世忠、Norman Cherry（英国）、任进、潘团结、齐红、戚鸟定、宋处岭、施堃、孙仲鸣、唐绪祥、滕菲、王春利、王春刚、王晓昕、王鼐、汪正虹、文乾刚、吴峰华、许平、许梦佳、月文、张纯辉、张福文、张莉君、张铁成、张世忠、赵丹绮、郑静、郑裕彤、郑耿坚、钟连盛、周桃林、周厚厚、周宗文、庄冬冬、邹宁馨

执行委员：

高伟、胡俊、傅永和、潘峰、赵祎、宋懿、熊芏芏、唐天、程之璐、刘小奇、王涛、韩欣然、王浩睿

Chairmen:

Ma Shengjie (Secretary of Party Committee of the Beijing Institute of Fashion Technology)
Luo Min (Director of the Industrial Culture Development Center of Ministry of Industry and Information Technology)

Vice Secretary General:

Ma Fengbao, Lan Cuiqin, Chang Wei

Members:

Ann Priest (UK), Bai Jingyi, Bao Xiaoying, Chu Weimin, Cheng Xuelin, Chen Guozhen, Chen Xiaohua, Eimera Conyard (Ireland), Guo Qiang, Guo Yingjie, Guo Xin, Gao Wei, Hong Xingyu, Hans Stofer (UK), Hu Shugang, Huang Xiaowang, Huang Wen, Lauret-Max De Cock (Belgium), Leo Caballero (Spain), Li Yingjie, Li Xuemei, Liao Chuangbin, Ma Shizhong, Norman Cherry (UK), Ren Jin, Pan Tuanjie, Qi Hong, Qi Niaoding, Song Chuling, Shi Kun, Sun Zhongming, Tang Xuxiang, Teng Fei, Wang Chunli, Wang Chungang, Wang Xiaoxin, Wang Nai, Wang Zhenghong, Wen Qiangang, Wu Fenghua, Xu Ping, Xu Mengjia, Yue Wen, Zhang Chunhui, Zhang Fuwen, Zhang Lijun, Zhang Tiecheng, Zhang Shizhong, Zhao Danqi, Zheng Jing, Zheng Yutong, Zheng Gengjian, Zhong Liansheng, Zhou Taolin, Zhou Houhou, Zhou Zongwen, Zhuang Dongdong, Zou Ningxin

---

Executive Member:

Gao Wei, Hu Jun, Fu Yonghe, Pan Feng, Zhao Yi, Song Yi, Xiong Dudu, Tang Tian, Cheng Zhilu, Liu Xiaoqi, Wang Tao, Han Xinran, Wang Haorui

INTEGRATION & CONNECTION

北 京 国 际 首 饰 艺 术 展 · 贯 通

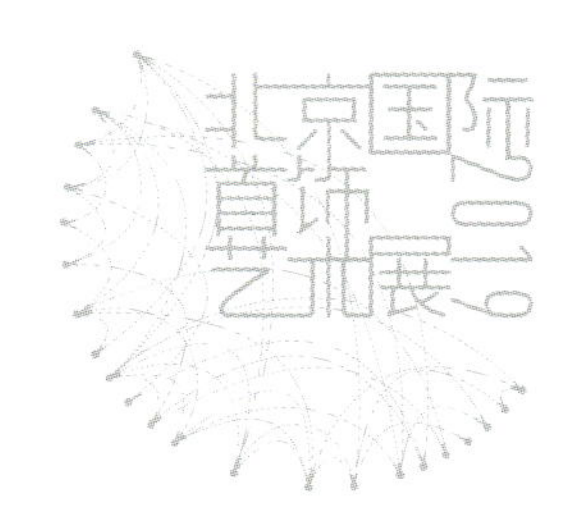

序言

2019 北京国际首饰艺术展

在人类文明的历史长河中，首饰一直伴随人类的美好生活而发展。随着生产力与生产技术的进步，在多元化发展的今天，首饰不仅是提高生活品质的重要艺术媒介，也是文化的载体，传播着一个民族的、人文的、地域的、历史的、自然的、环境的、人性的内涵，折射出思想、期盼、愿望以及情感。首饰是佩戴者张扬个性、展示自我的标志，为人们的生活增加了情感、美感与诗意。

秋天是北京最美好的季节。2 年前，2017 北京国际首饰艺术展在北京服装学院隆重举行，来自全球 40 多个国家和地区的 300 多位首饰艺术家共同倡议发起国际首饰设计高校联盟。经过两年多的积极筹备，通过多方的努力，“国际首饰设计高校联盟”的筹建工作终于有了成果。

作为一所国际化的时尚院校，北京服装学院发起并主持筹建国际首饰设计高校联盟，符合将学校建设成为国际一流时尚院校的愿景，同时也符合中国在全球化背景下，向世界展示开放、合作的态度。首饰设计是人类美好生活构建的重要组成部分，国际首饰设计高校联盟的建立将聚集全球优势资源，把握时代脉搏，通过设计发声，以首饰设计为媒介，促进全球各国之间文化交流、教育协同，满足人们对美好生活的向往。

国际首饰设计高校联盟的成立，将定义首饰设计、首饰技术的最高标准。以服务为宗旨，以自愿、平等、合作、共赢为原则，面向国际，服务首饰产业，开放创新，协作发展。首饰联盟将紧跟国际设计产业发展趋势，坚持不断创新的理念，提升联盟群体竞争力，树立并维护联盟形象。充分发挥首饰联盟的优势，推进首饰设计教育的发展。为建立一个和谐、综合、可持续发展的国际首饰设计资源共享平台而努力。

预祝国际首饰设计高校联盟成立大会暨 2019 北京国际首饰艺术展圆满成功！

北京服装学院　校务委员会主席 马胜杰

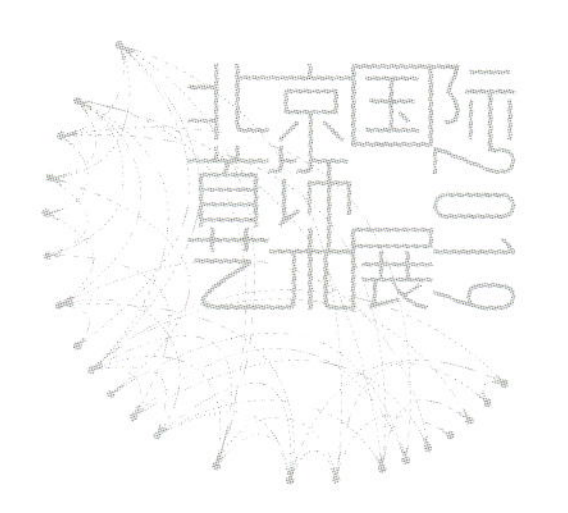

PREFACE

2019 BEIJING INTERNATIONAL JEWELLERY ART EXHIBITION

In the long history of human civilization, jewellery has been accompanying the improvement of human life. With the development of productive forces and production technology, jewellery is not only an important artistic medium for improving life quality, but also a carrier of culture.It communicates the connotation of nations, culture, regions, history, nature, environment and humanity, and reflects thoughts, expectations, desires and emotions. Jewellery is a symbol of the wearer's personality and self-expression, adding emotion, beauty and poetry to people's life.

Autumn is the best season in Beijing. Two years ago, the 2017 Beijing International Jewellery Art Exhibition was held in Beijing Institute of Fashion Technology(BIFT). More than 300 jewellery artists from more than 40 countries and regions initiated the International Jewellery College Association. After more than two years of efforts made by various parties, the preparatory work of the "International Jewellery College Association" has finally come to fruition.

As an International Fashion College, BIFT initiated and presided over the establishment of the International Jewellery College Association. This act is in line with the vision of building the school into an international first-class fashion college, and also in line with China's desire to display to the world an open and cooperative attitude against the globalization background. Jewellery design is an important part of constructing a better life for human beings. The establishment of the International Jewellery College Association will pool the global advantageous resources, take the pulse of the times, make a sound through design, and take jewellery design as a medium to promote cultural exchanges and educational synergy among countries around the world to meet people's aspirations for a better life.

The establishment of the International Jewellery College Association will set the highest standards for jewellery design and jewellery technology. Take services as the tenet, take "voluntary, equal, cooperative and win-win" as the principle, the association targets the International service jewellery industry and promotes open innovation and coordinated development. It will follow the development trend of the international design industry, adhere to the concept of continuous innovation, enhance the competitiveness of the association, establish and maintain the image of the association. It will also give full play to the advantages of the jewellery association to promote the development of jewellery design education. The association will strive to establish itself as a harmonious, comprehensive and sustainable international jewellery design resource sharing platform.

I wish the 2019 Beijing International Jewellery Art Exhibition and the Founding Conference of the International Jewellery College Association a complete success!

Mr. Ma Shengjie, Chairman of the Council, BIFT

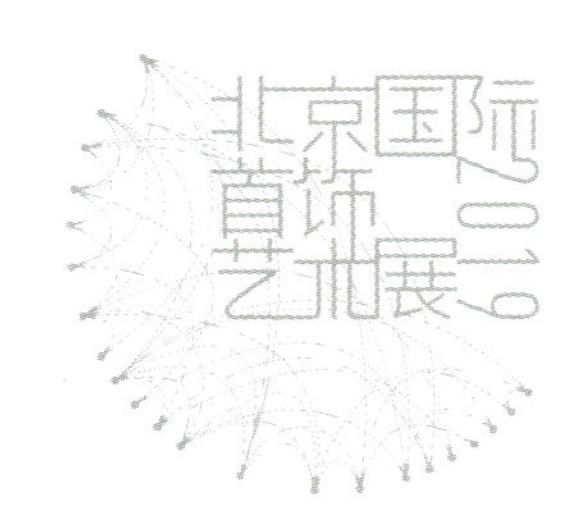

前言

2019 北京国际首饰艺术展

北京的十月，是金秋的时节、丰收的时节、奉献的时节。此时此刻，2019 北京国际首饰艺术展的展厅里，展示着来自世界 40 多个国家和地区、300 多位艺术家和设计师、超过一千件的首饰艺术作品。这是一场名副其实的当代首饰艺术饕餮盛宴，其展出作品质量之高、参展人数之多，都是世所罕见的。

事实上，作为两年举办一次的“双年展”，我们自 2013 年起，已经成功地举办四届。当今的时代，大众消费文化影响着我们日常生活的方方面面，作为“地球村”文化的幕后推手，它推动着“全球化”的浪潮滚滚向前，使“扁平化”的时代特征日趋明显。不过，就在“娱乐至死”的大众消费文化的边缘，精英文化茕茕孑立，顽强地守望着我们的精神家园。而当代首饰艺术作为精英文化一员，以其大胆创新的艺术实践，扮演了大众文化与时尚文化启蒙先锋的角色，引导首饰市场推陈出新，由此引发了越来越多的大众消费者的关注。

应该说，当代首饰艺术创作越来越多元，创作者的性情、素养、价值观以及生活经历也是多种多样，作品的风格也就各有千秋。求同存异，这是社会开放之后才会有的文化胸襟；价值多元，这是海纳百川的必然结果；而“贯 通”，也一定是全球文化走向共同繁荣的基石。只要我们笃定初心、砥砺前行，就一定会再次收获丰硕的文化艺术成果，就一定会再有丰收的季节、奉献的季节。

预祝 2019 北京国际首饰艺术展成功举办！

北京服装学院　副校长 詹炳宏

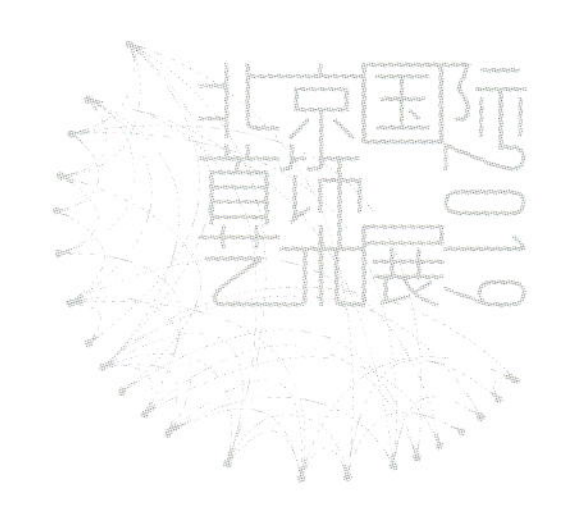

FOREWORD

2019 BEIJING INTERNATIONAL JEWELLERY ART EXHIBITION

October is the golden autumn season for harvest and dedication. At this moment, 2019 Beijing International Jewellery Art Exhibition is showcasing more than 1000 pieces of jewellery art works by more than 300 artists and designers from more than 40 countries and regions. This is indeed a feast of contemporary jewellery and art. The high quality of the exhibited works and the number of attendees are rarely in the world.

Since 2013, this biennial exhibition has been successfully held for four times. Nowadays, mass consumer culture, the hand behind the "global village" culture, imposes an impact on every aspect of our daily life, advancing the wave of "globalization" and highlighting the "flattening" characteristic of our times. However, elite culture, standing alone in the edge of the mass consumer culture which gives priority to entertainment, still keeps watching over our spiritual home. Boasting its audacious art practice, contemporary jewellery art, a member of elite culture, has been a pioneer in the enlightenment of mass culture and fashion culture, leading the innovation of the jewellery market and thus attracting more attention from mass consumers.

It should be mentioned that the creation of the contemporary jewellery art is increasingly diversified as the style of each item has its own merits due to the diversification of creators' temperament, accomplishment, value and experience. The cultural mindset of seeking common ground while preserving differences was developed after the opening-up policy; value pluralism is an inevitable result of being tolerant to diversity; while "communication and exchange" is the basis for global culture to develop towards common prosperity. As long as we forge ahead while remaining true to the original aspiration, we will acquire abundant achievements of culture and art, and embrace a season of harvest and dedication again.

I wish the 2019 Beijing International Jewellery Art Exhibition a great success!

Zhan Binghong, Vice President,
Beijing Institute of Fashion Technology

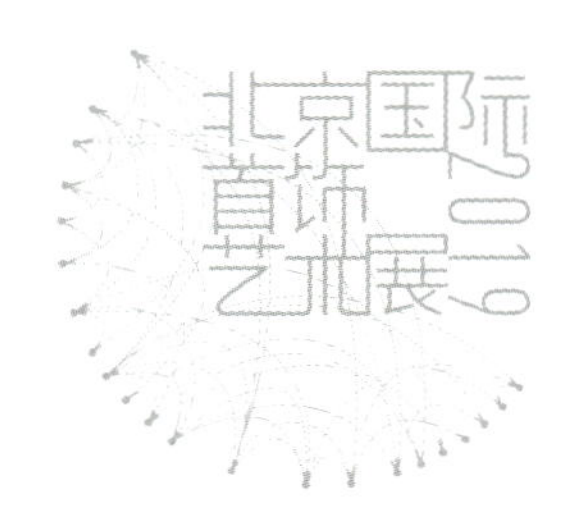

致辞

2019 北京国际首饰艺术展

我很荣幸代表国际设计协会 (ico-D) 欢迎您参加本次活动。ico-D 创建于 1963 年，是设计行业最大的国际组织，由国家级专业设计师组织、设计院校和设计推广机构组成。

我们很高兴北京服装学院成为我们的会员。作为中国领先的时尚设计院校，北京服装学院为 ico-D 和我们所有的国际成员提供了一个与充满活力的中国经济互动的渠道。

或许更重要的是，北京服装学院为我们提供了洞察中国丰富的遗产和传统以及极其重要的未来中国国内市场的机会。这个市场不仅将成为全球经济的引擎，还将决定国际文化和消费趋势——包括时尚趋势。作为一个国际性的设计组织，我们非常支持北京服装学院关于创建国际首饰设计高校联盟的想法。

为了迎接 21 世纪和全球化的挑战，设计专业正在经历一个历史性的变革时期。 这意味着，整个设计界——从业人员和教育工作者——必须共同努力，重新定义什么是专业设计师，当然，这意味着设计课程也必须演变发展——包括首饰设计课程。首饰设计蕴含文化、技术、艺术等诸多因素，在世界各国都有所发展。2019 北京国际艺术首饰展作为世界首饰艺术沟通交流的桥梁，将为首饰艺术的发展起到不可忽视的作用。感谢北京服装学院邀请我参加这次活动，我衷心地祝贺国际首饰设计高校联盟成立大会暨 2019 北京国际首饰艺术展圆满成功。

国际设计联合会前主席　大卫 · 古施曼

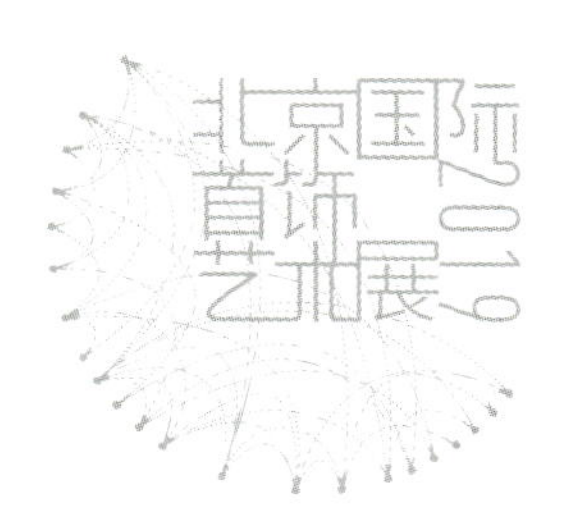

SPEECH

2019 BEIJING INTERNATIONAL JEWELLERY ART EXHIBITION

On behalf of ico-D, the International Council of Design, it is my honor to welcome you to participate in this event. ico-D, established in 1963, is the largest international organization representing the professions of design, composed of national professional designer organizations, design schools and design promotion entities.

We are happy to count the Beijing Institute of Fashion Technology among our Members. BIFT, as the leading fashion design school in China, provides ico-D and all our international members a channel for interacting with the dynamic Chinese economy, and perhaps more importantly offers us insight into both China' s rich heritage and traditions and the hugely important future internal Chinese market. That market, will not only be the engine of the global economy, but will determine international cultural and consumption trends - including fashion trends.

As an international design organization, we are also very supportive of the idea of the International Jewellery College Association. The design professions are going through a period of historic change as they evolve to meet the challenges of the 21st century and Globalization. This means that the entire design community - practitioners and educators - must work together to redefine what it means to be a professional designer, and of course this means that design curricula must also evolve - including the curricula of jewellery design programs. There are multiple element had been involved by jewellery design including culture ,technology and arts,and had been developed all over the world , 2019 Beijing International Art jewellery exhibition as the bridege between international art jewellery that will imporving the development of jewellery as art and design subjects. ico-D wishes the International Jewellery College Association success in its meetings here in Beijing, and witsh 2019 Beijing international Art jewellery exhibition successful. Thanks for inviting me to attend this event and wish you all success.

David Grossman ,Past President of ico-D,
International Council of Design (ico-D)

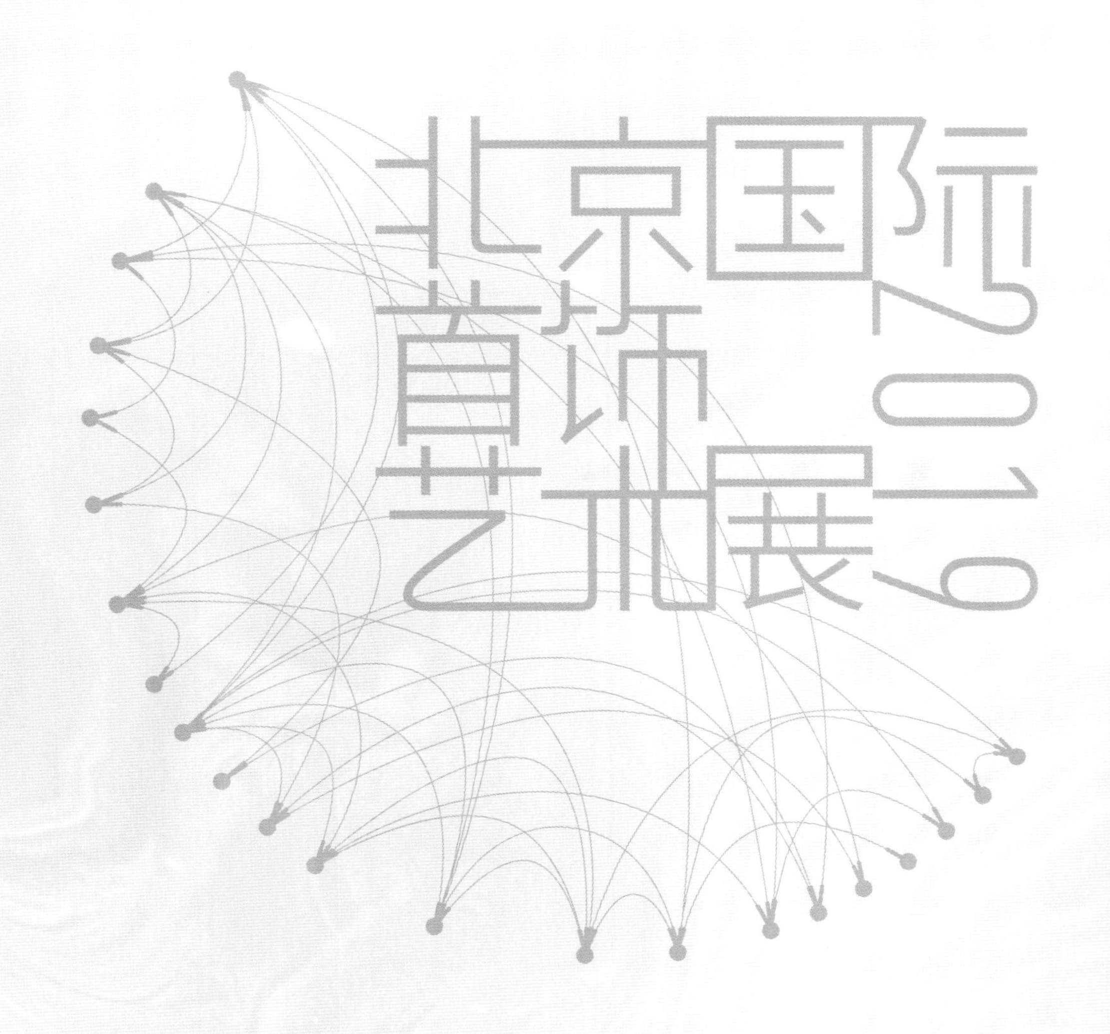

目 录
CONTENTS

ON THE

EXHIBITI

展览现场

ON

北京国际首饰艺术展2019

BIFT,FASHION CHINA
SCHOOL OF FASHION
SCHOOL OF FASHION
ACCESSORY
ART & ENGINEERING

国际首饰设计高校联盟成立大会暨
2019北京国际首饰艺术展
Founding conference of The international jewelry colleges association and
2019 Beijing International Jewelry Art Exhibition

首饰设计高校联盟成立大会
9北京国际首饰艺术展开幕式
nference of The international jewelry colleges association and
emony of 2019 Beijing International Jewelry Art Exhibition
10月18日下午 14:30开幕仪式签到
装学院 服饰艺术与工程学院
n Accessory Art & Engineering, B
f Fashion Technology
ber, 2019)

国际首饰设计高校联盟成立大会
暨2019北京国际首饰艺术展开幕式
Founding conference of The international jewelry colleges association and
Opening ceremony of 2019 Beijing International Jewelry Art Exhibition

OPENING SPEECH
开幕致辞
David Gros   nan
国际设计联   席
Past President of the In   esign

OPENING SPEECH
开幕致辞
马胜杰
北京服装学院

BTV

国际首饰设计高校联盟成立大会
暨2019北京国际首饰艺术展开幕秀
Founding Conference of The International Jewellery College Association
& Opening Ceremony of 2019 Beijing International Jewellery Art Exhibition

国际首饰设计高校联盟成立大会
暨2019北京国际首饰艺术展开幕秀
Founding Conference of The International Jewellery College Association
& Opening Ceremony of 2019 Beijing International Jewellery Art Exhibition

国际首饰设计高校联盟成立大会暨
2019北京国际首饰艺术展
Founding conference of The international jewelry colleges association and
2019 Beijing International Jewelry Art Exhibition

演讲嘉宾

Awards

2019国际首饰艺术高峰论坛

国际首饰设计高校联盟
International Jewellery College Association
2019国际首饰设计高校教育座谈会
International symposium for jewellery design in higher education

LIST OF ARTISTS

参展艺术家

北京国际
首饰
艺术展
2019

特邀参展艺术家作品 | WORKS OF INVITED ARTISTS

外国参展艺术家作品 | WORKS OF OVERSEA ARTISTS

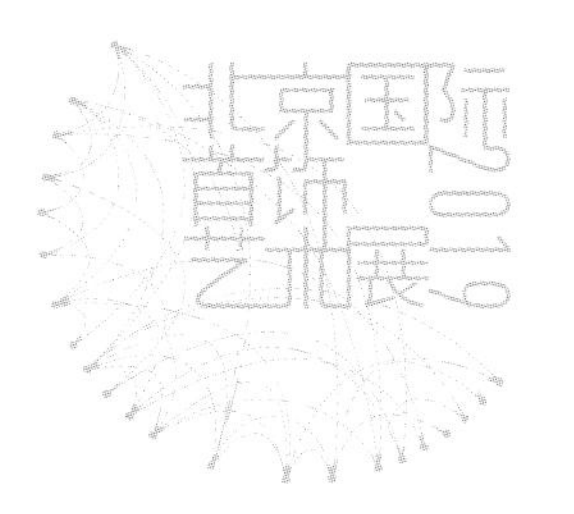
北京国际
首饰
艺术展
2019

中国参展艺术家作品 | WORKS OF CHINESE ARTISTS

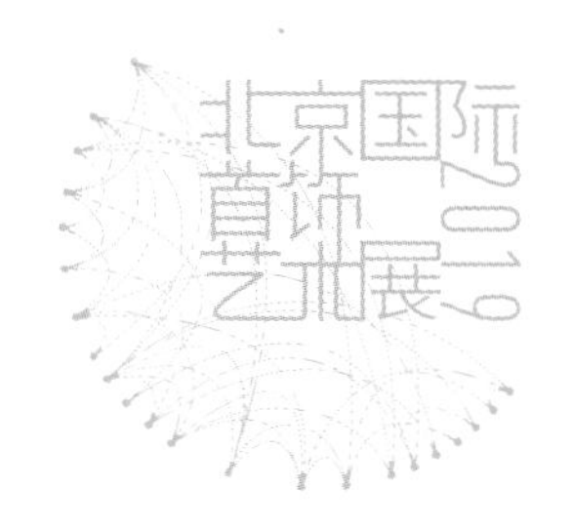
北京国际
首饰
艺术展
2019

WORKS OF

INVITED ARTISTS

特邀参展艺术家作品

并不是全部、超出你的想象
NOT ALL IT APPEARS TO BE, MORE THAN YOU CAN IMAGINE

作者姓名：Chris De Beer（南非）
作品类型：胸针
作品材质：聚丙烯，安全别针等

Artist: Chris De Beer (South Africa)
Type: brooch
Material: polypropylene, safety pin, etc.

“跳”2018
“JUMP”2018

作者姓名：Shirly Bar Amotz（以色列）
作品类型：项链
作品材质：铜，铁，热珐琅，钻石粉末，环氧树脂腻子

Artist: Shirly Bar Amotz（Israel）
Type: necklace
Material: copper, iron, hot enamel, diamond powder, epoxy resin putty

现象
PHENOMENON

作者姓名：Hiroki Iwata（日本）
作品类型：胸针
作品材质：银

Artist: Hiroki Iwata（Japan）
Type: brooch
Material: silver

红玫瑰
ROSA RUBEA HOMUNCULA

作者姓名：Jivan Astfalck（英国 / 德国）
作品类型：项饰
作品材质：925 银，粉末涂层，黄铜，压钉

Artist:	Jivan Astfalck（Britain / Germany）
Type:	necklace
Material:	sterling silver, powder coated, brass, press studs

给它一个标记，黄色的印记
MARK-IT , YELLOW-TRACES

作者姓名：Stephen Bottomley（英国）
作品类型：胸针
作品材质：银，18k 黄金，亚克力，珐琅钢，橡胶

Artist: Stephen Bottomley (Britain)
Type: brooch
Material: silver, 18ct gold, acrylic, enamel steel, rubber

梦幻套房
DREAM SUITE

作者姓名：Tom Muir（英国）
作品类型：胸针
作品材质：银，18k 黄金，亚克力，珐琅钢，橡胶

Artist: Tom Muir (Britain)
Type: brooch
Material: silver, 18ct gold, acrylic, enamel steel, rubber

由钯金箔细节装饰的白色浅灰色条纹臂饰
ARMPIECE WHITE LIGHT GREY STRIPE, PALLADIUM LEAF DETAIL

作者姓名：Angela O'Kelly（爱尔兰）
作品类型：臂饰
作品材质：纸，毛毡，钯金箔，弹性纤维

Artist: Angela O'kelly（Ireland）
Type: bracelet
Material: paper, felt, palladium leaf, elastic

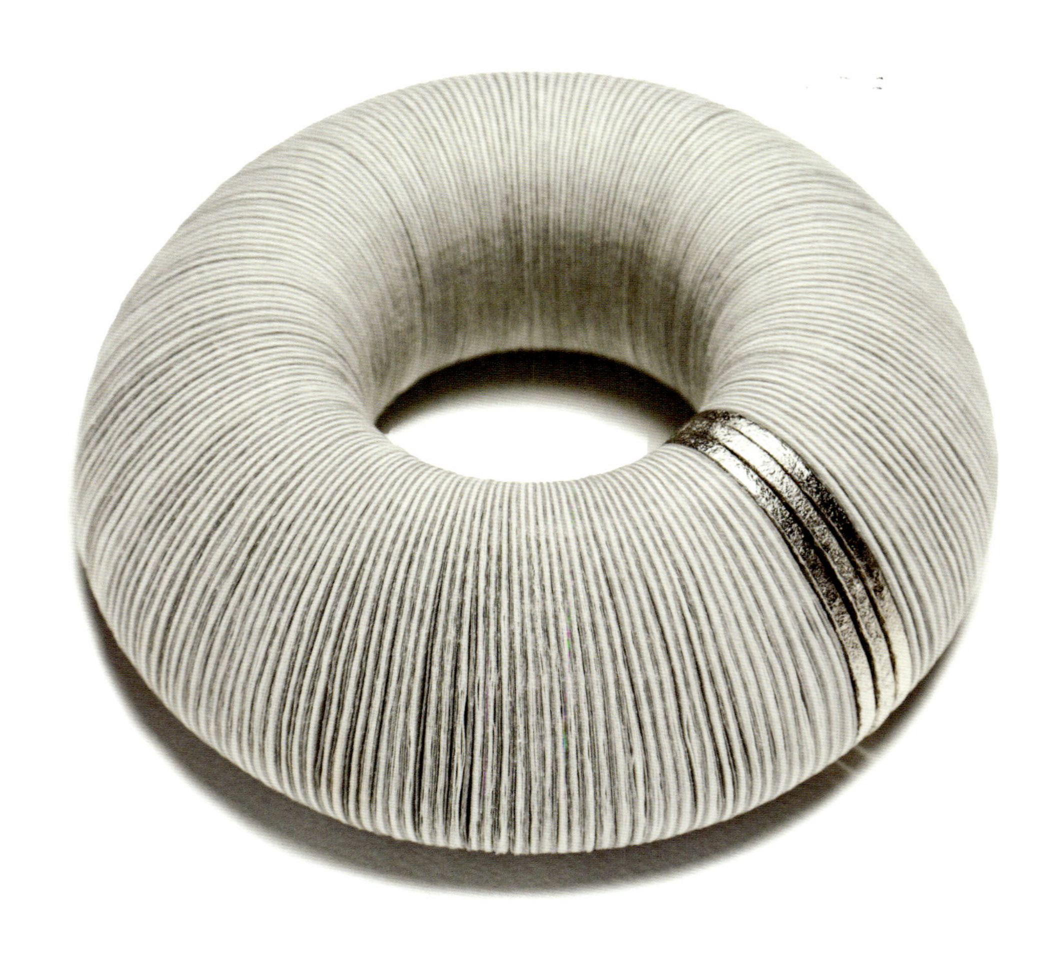

WORKS OF

OVERSEA ARTISTS

外国参展艺术家作品

宁菲亚
NINFEA

作者姓名：Adelina Mummolo（意大利）
作品类型：手镯
作品材质：银，铜锈

Artist: Adelina Mummolo (Italy)
Type: bracelet
Material: silver, patina

大O红石
THE BIG O RED ROCK

作者姓名：Adriana Almeida Meza （哥伦比亚）
作品类型：项链，胸针
作品材质：岩石水晶，天然漆，银，钢

Artist: Adriana Almeida Meza (Colombia)
Type: necklace , brooch
Material: rock crystal, lacquer, silver, steel

在结束前
BEFORE THE END

作者姓名：Alice Balestro Floriano（巴西）
作品类型：耳环
作品材质：银，PLA 塑料和木头

Artist: Alice Balestro Floriano（Brazil）
Type: earring
Material: silver, pla plastic and wood

桃花
PEACH FLOWER

作者姓名：Alice Biolo（意大利）
作品类型：项链
作品材质：银，氟涂料

Artist: Alice Biolo（Italy）
Type: necklace
Material: silver, fluoropolymer paints pigment

你好
HELLO

作者姓名：Andrea Auerl（奥地利）
作品类型：项链
作品材质：古董电木话筒，铅珠，玻璃水晶，电缆，不锈钢

Artist: Andrea Auerl (Austria)
Type: necklace
Material: bakelite mouthpiece of an antique phone, lead beads, glass crystal, electric cable, stainless steel

长路
A LONG ROAD

作者姓名：Andrea Gambato（意大利）
作品类型：项链
作品材质：银，铜，铜锈

Artist: Andrea Gambato (Italy)
Type: necklace
Material: silver, copper, patina

我等移民
WE THE MIGRANTS

作者姓名：Andrea Rosales-Balcarcel（危地马拉）
作品类型：胸针
作品材质：树脂，水彩颜料，彩铅，18k 金，钢等

Artist: Andrea Rosales-Balcarcel（Guatemala）
Type: brooch
Material: resin, watercolor, color pencil, 18ct gold, steel, etc.

生活，2019 年
LEBEWESEN ,2019

作者姓名：Anna Babett Von Dohnanyi（德国）
作品类型：项饰
作品材质：青铜

Artist: Anna Babett Von Dohnanyi（Germany）
Type: necklace
Material: bronze

丢失的牙齿
LOST TEETH

作者姓名：Anne Luz Castellanos（墨西哥）
作品类型：胸针，项链
作品材质：钢琴键，银等

Artist: Anne Luz Castellanos（Mexico）
Type: brooch, necklace
Material: piano key, silver, etc.

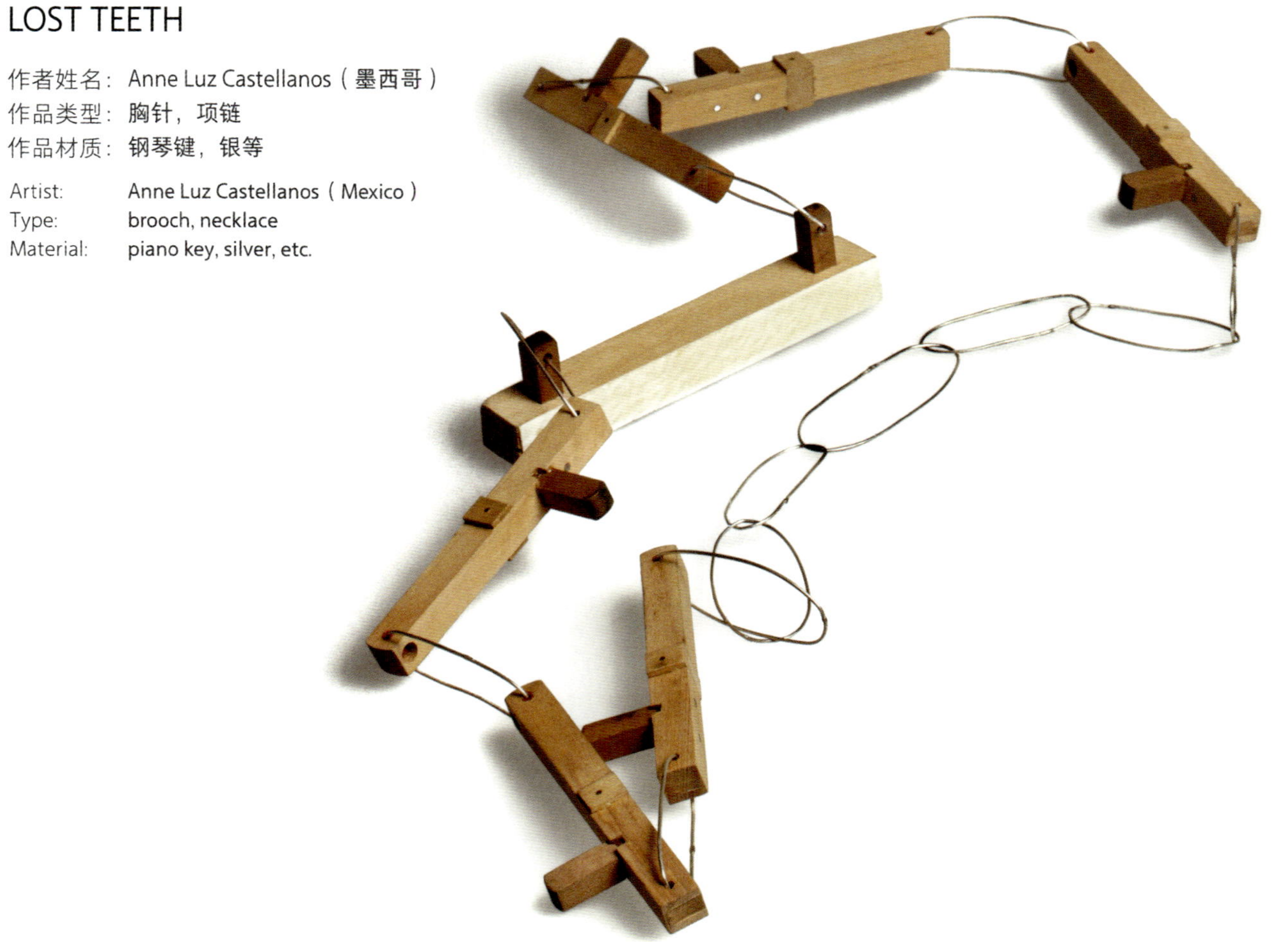

来自天堂的信息
MESSAGE FROM THE HEAVEN

作者姓名：Arang Kim（阿曼）
作品类型：胸针，饰针，项链
作品材质：925 银，葡萄石

Artist: Arang Kim（Oman）
Type: brooch, necklace
Material: sterling silver, prehnite

“带我回去！”
“TAKE ME BACK！”

作者姓名：Arianne Kresandini（印度尼西亚）
作品类型：胸针
作品材质：银，黄铜，铜，红色的线，树脂，烟草

Artist: Arianne Kresandini（Indonesia）
Type: brooch
Material: silver, brass, copper, red thread, resin, tobacco

两者之间 YOBUN 绿
IN BETWEEN YOBUN GREEN

作者姓名：Bin Shehab Saadah（也门）
作品类型：戒指，胸针，项链
作品材质：宝石，黄铜，银，木头，乙烯基等

Artist: Bin Shehab Saadah（Yemen）
Type: ring, brooch, necklace
Material: gemstone, brass, silver, wood, vinyl, etc.

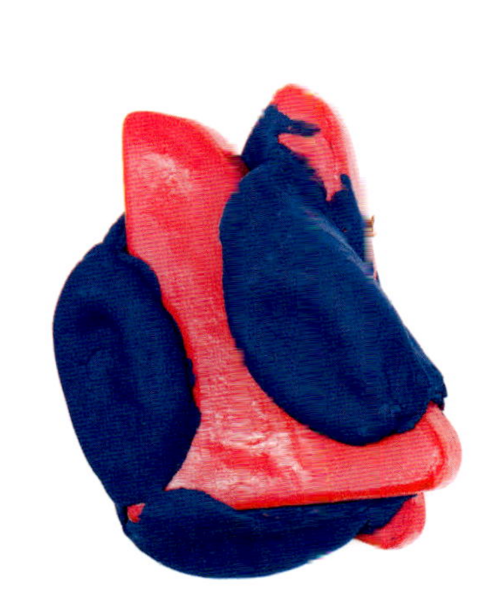

光与影体宝石，胸部项链
LIGHT AND SHADOW BODYJEWEL BOOBS NECKLACE

作者姓名：Caterina Zucchi（意大利）
作品类型：项链
作品材质：穆拉诺玻璃

Artist: Caterina Zucchi (Italy)
Type: necklace
Material: murano glass

穆拉诺人世窗，人类灭绝的花，穆拉诺的风景
ANTHROPOCENE WINDOWS
ANTHROPOCENE EXTINCT FLOWERS
ANTHROPOCENE LOST LANDSCAPES

作者姓名：Chiara Scarpitti（意大利）
作品类型：项链，胸针，饰针
作品材质：银铑，黑钢，真丝，有机玻璃

Artist: Chiara Scarpitti (Italy)
Type: necklace, brooch
Material: silver rhodium, black steel, pure silk, plexiglass

你好
"Hod ó s"

作者姓名：Claire Lavendhomme（比利时）
作品类型：胸针
作品材质：钢，金

Artist: Claire Lavendhomme（Belgium）
Type: brooch
Material: steel, gold

托斯卡纳景观，黎明
托斯卡纳风景，日落
TUSCAN LANDSCAPE, DAWN
TUSCAN LANDSCAPE, SUNSET

作者姓名：Corrado De Meo（意大利）
作品类型：胸针，饰针
作品材质：银，纸板，丙烯酸金属漆，氧化物，铁

Artist: Corrado De Meo（Italy）
Type: brooch
Material: silver, cardboard, acrylic metallic paint, oxides, iron

我的朋友，请把你的手给我
兄弟我们一起走

GIVE ME YOUR HAND MY FRIEND
WE WILL WALK TOGETHER, BROTHER

作者姓名：Daniel von Weinberger（比利时）
作品类型：项饰
作品材质：塑料

Artist: Daniel von Weinberger (Belgium)
Type: necklace
Material: plastic

LUMINA 系列胸针
LUMINA SERIES BROOCH

作者姓名：Donald Friedlich（美国）
作品类型：胸针
作品材质：14K 金，二色玻璃，硼砂酸盐玻璃等

Artist: Donald Friedlich (U.S.A)
Type: brooch
Material: 14ct gold, dichroic glass, borosilicate glass, etc.

指尖，领口和袖口
FINGERTIPS, COLLAR AND CUFFS

作者姓名：Dr Marlene De Beer（南非）
作品类型：其他
作品材质：925 银和棉花，钢

Artist: Dr Marlene De Beer（South Africa）
Type: other
Material: sterling silver and cotton, steel

平衡戒指，天门戒指
EQUILIBRIO RING , TIANMEN RING

作者姓名：Ebrahim Mohammadian Elird（波黑）
作品类型：戒指
作品材质：银，珐琅，油漆，石英

Artist: Ebrahim Mohammadian Elird（Bosnia and Herzegovina）
Type: ring
Material: silver, enamel, oil paint, quartz

飞翔的猛兽
FLYING BEAST

作者姓名：Egle Sitkauskaite（立陶宛）
作品类型：胸针
作品材质：925 银，被柠檬酸处理过的紫铜，钢

Artist: Egle Sitkauskaite（Lithuania）
Type: brooch
Material: sterling silver, cooper citrin, steel

致祖父：过去即现在，
致祖母：过去即现在，
致她：过去即现在
PAST IS PRESENT (GRANDFATHER),
PAST IS PRESENT (GRANDMOTHER),
PAST IS PRESENT (HER)

作者姓名：Elina Honkanen（芬兰）
作品类型：胸针
作品材质：聚碳酸酯，银，钢，银箔

Artist: Elina Honkanen（Finland）
Type: brooch
Material: polycarbonate plates, silver, steel, silver leaf

无标题
UNTITLE

作者姓名：Elli Xippa（希腊）
作品类型：项链
作品材质：连裤袜，胶水，金属线，铜

Artist: Elli Xippa（Greece）
Type: necklace
Material: pantyhose, glue, wire, copper

成为，打破，出现
BECOME，BREAKOUT，COME OUT

作者姓名：Elwy Schutten（荷兰）
作品类型：项链
作品材质：银，皂石

Artist: Elwy Schutten（Netherlands）
Type: necklace
Material: silver, soapstone

水晶欧泊
OPAL CRYSTAL

作者姓名：Emil Weis Opals（德国）
作品类型：胸针
作品材质：18k 金，钻石

Artist: Emil Weis Opals（Germany）
Type: brooch
Material: 18ct gold, diamond

蜻蜓项链，泽曼项链，圣甲虫魔杖项链
DRAGONFLY NECKLACE , ZEMAN NECKLACE SCARAB MAGIC WAND NECKLACE

作者姓名：Ena Mulavdic（波黑）
作品类型：项饰
作品材质：核桃木，血木，椰木，无花果木，玻璃，铁，珐琅，油漆，丙烯酸漆，黏土，赤铁矿，绳子

Artist: Ena Mulavdic（Bosnia and Herzegovina）
Type: necklace
Material: walnut wood, blood wood, coconut wood, fig wood, glass,iron, enamel, paint, acrylic paint, clay, hematite, cord

模仿者，我
COPYCATS I

作者姓名：Erato Kouloubi（希腊）
作品类型：胸针
作品材质：意大利面，丙烯酸，青铜

Artist: Erato Kouloubi（Greece）
Type: brooch
Material: pasta, acrylics, bronze

繁荣，漂浮，原始
FLOURISH , AFLOAT , RAW

作者姓名：Eva Fernandez（西班牙）
作品类型：胸针
作品材质：铝，不锈钢，木材，钢

Artist: Eva Fernandez（Spain）
Type: brooch
Material: aluminium, stainless steel, wood, steel

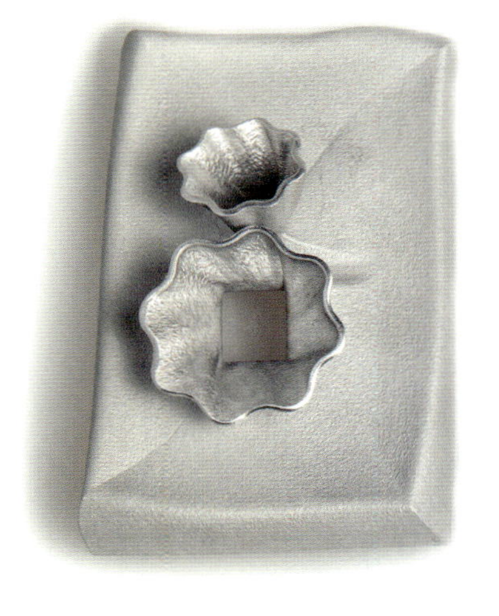

星球，轨道
PLANETS , RAIL

作者姓名：Evgenia Zoidaki（希腊）
作品类型：手镯，戒指
作品材质：纯银 925，钢垫，钢丝

Artist: Evgenia Zoidaki（Greece）
Type: bracelet, ring
Material: sterling silver 925, mat, wires of steel

铁网花
STEEL MESH FLOWER

作者姓名：Fumie Sasaki（日本）
作品类型：胸针
作品材质：银，不锈钢，金属网，油漆

Artist: Fumie Sasaki（Japan）
Type: brooch
Material: sliver, stainless, steelmesh, paint

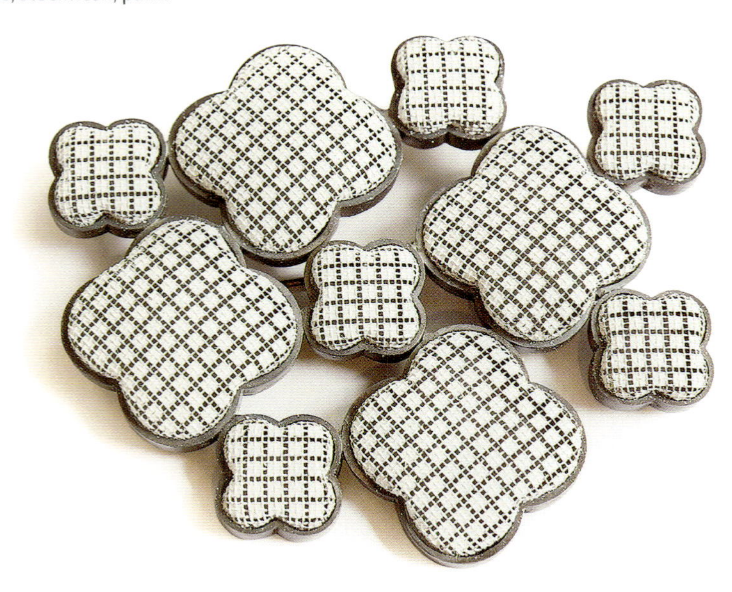

空气
AIR

作者姓名：Gabriella Vincze-Baba（匈牙利）
作品类型：戒指，耳环，项链
作品材质：925 银，薄纱

Artist: Gabriella Vincze-Baba (Hungary)
Type: ring, earring, necklace
Material: sterling silver, tulle

独自一人，文本意境，框架绘画意境，黑色，绘画意境系列
ALONE, TEXT SERIE , FRAME PAINTING SERIE , BLACK, PAINTING SERIE

作者姓名：Gigi Mariani（意大利）
作品类型：胸针，饰针
作品材质：银，18k 黄金，乌银，铜锈

Artist: Gigi Mariani (Italy)
Type: brooch
Material: silver, 18ct gold, niello, patina

切割分支，红包耳环
CUTTING BRANCHES, HONG BAO EARRINGS

作者姓名：Gussie van der Merwe（南非）
作品类型：项链，耳环，戒指
作品材质：铜，银，仿金箔，线

Artist: Gussie van der Merwe（South Africa）
Type: necklace, earrings, ring
Material: copper, silver, imitation gold leaf, thread

一个我想知道的故事，双轮自行车，可爱的朋友
A STORY I WANT TO KNOW，TWO-WHEELED BICYCLE，LOVELY FRIENDS

作者姓名：Han Soon In（韩国）
作品类型：胸针，饰针，项链
作品材质：925 银，白、黑珍珠，黑钻石

Artist: Han Soon In（South Korea）
Type: brooch, necklace
Material: sterling silver, white & black pearl, black diamond

分享
PARTAKE

作者姓名：Hayan Kim（韩国）
作品类型：项链
作品材质：丙烯酸玻璃，不锈钢

Artist: Hayan Kim（South Korea）
Type: necklace
Material: acylic glass, stainless steel

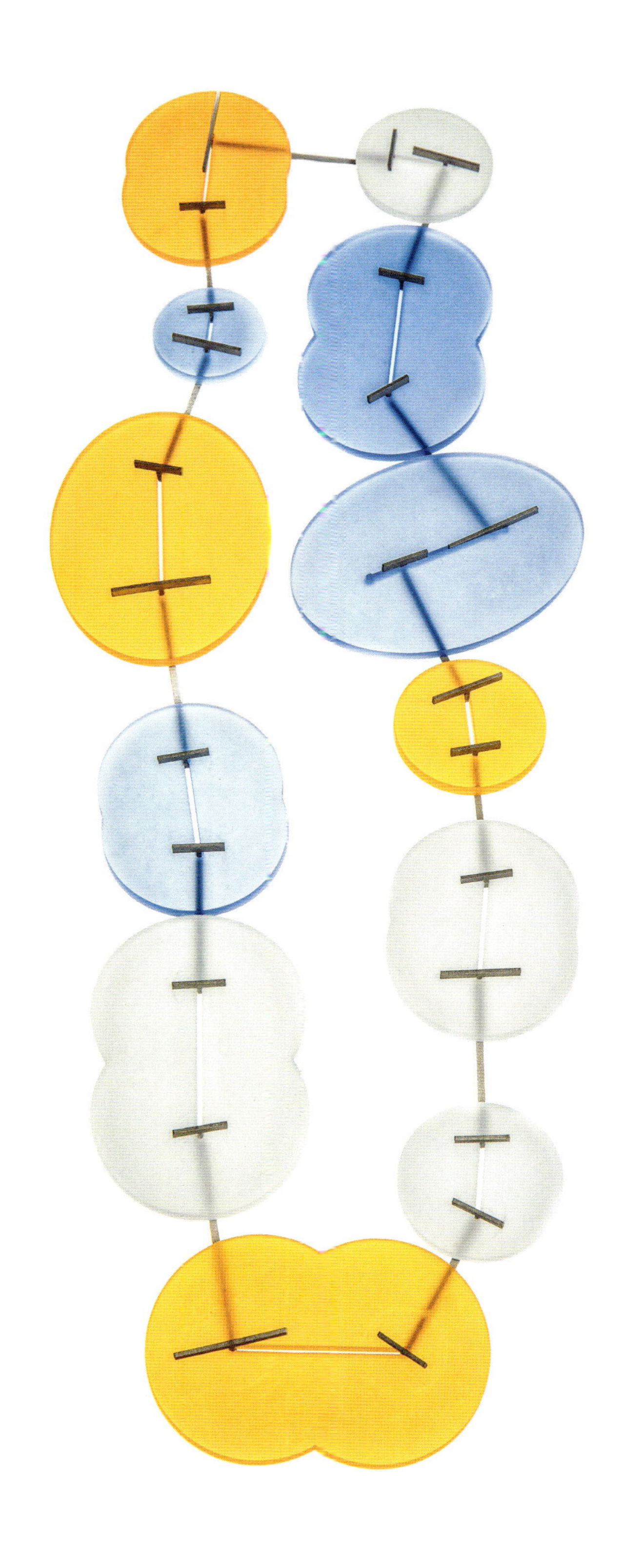

无题（三套）
UNTITLE

作者姓名：Hirte Lydia（德国）
作品类型：项链
作品材质：布里斯托尔板，照片卡，书法墨水，釉面木，珠丝

Artist: Hirte Lydia（Germany）
Type: necklace
Material: bristol board, photo card, calligraphic ink, wood glaze, beading silk

午夜后系列
AFTER MIDNIGHT

作者姓名：Ildiko Juhasz（新西兰）
作品类型：胸针
作品材质：银，丙烯，电线等

Artist: Ildiko Juhasz（New Zealand）
Type: brooch
Material: silver，acrylic paint，electriacal wire, etc.

新式领圈
NEW NECKLINES

作者姓名：Jana Graf（德国）
作品类型：项链
作品材质：塑形碳，珍珠丝，手染上蜡棉线

Artist: Jana Graf（Germany）
Type: necklace
Material: carbon (shaped), pearlsilk, cotton rope (hand-dyed, waxed)

“翼 II-5”花脊吊坠
"WINGS II-5" THE VERTEBRATION OF FLOWER

作者姓名：Jean-Marc Waszack（法国）
作品类型：胸针，项链
作品材质：钛金属，珍珠，植物纤维，醋酸纤维素

Artist: Jean-Marc Waszack（France）
Type: brooch, necklace
Material: titanimum, pearl, plant fiber, cellulose acetate

蓝色连接，纤细的冰，柔软的水
INTERCONNECTION OF BLUE THIN ICE, SOFT WATERS BLUE ROOTS AND BLACK ROOTS CONNECTION

作者姓名：Jekaterina Smirnova（拉脱维亚）
作品类型：胸针，其他，项链
作品材质：重组青金石，月亮石，黑玛瑙，马毛

Artist: Jekaterina Smirnova（Latvia）
Type: brooch, other, necklace
Material: reconstructed lapis, lazuli stone, moonstone, black onyx, horse hair

摩尔系列
MOLE SERIES

作者姓名：Jeong Ji Eun（韩国）
作品类型：胸针
作品材质：925 银，聚氨酯

Artist: Jeong Ji Eun（South Korea）
Type: brooch
Material: sterling silver, urethane

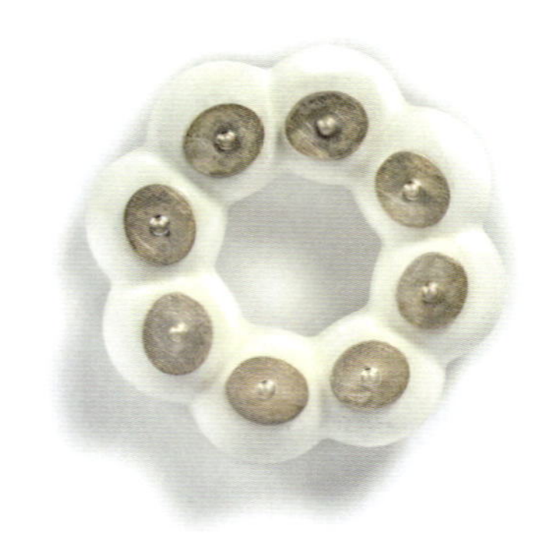

月球背面 21 系列
BACK OF THE MOON 21 SERIES

作者姓名：Ji Young Kim（韩国）
作品类型：胸针，项链
作品材质：925 银，镀金，钢

Artist: Ji Young Kim（South Korea）
Type: brooch, necklace
Material: sterling silver, gold-plated, steel

我是一个女人，第一晚，内而外
I'M A WOMAN, FIRST NIGHT, INSIDE OUT

作者姓名：Jieun Park（韩国）
作品类型：胸针，项链
作品材质：铜，钢，丙烯酸漆，棉线等

Artist: Jieun Park（South Korea）
Type: brooch, necklace
Material: brass, steel, acrylic lacquer, cotton thread, etc.

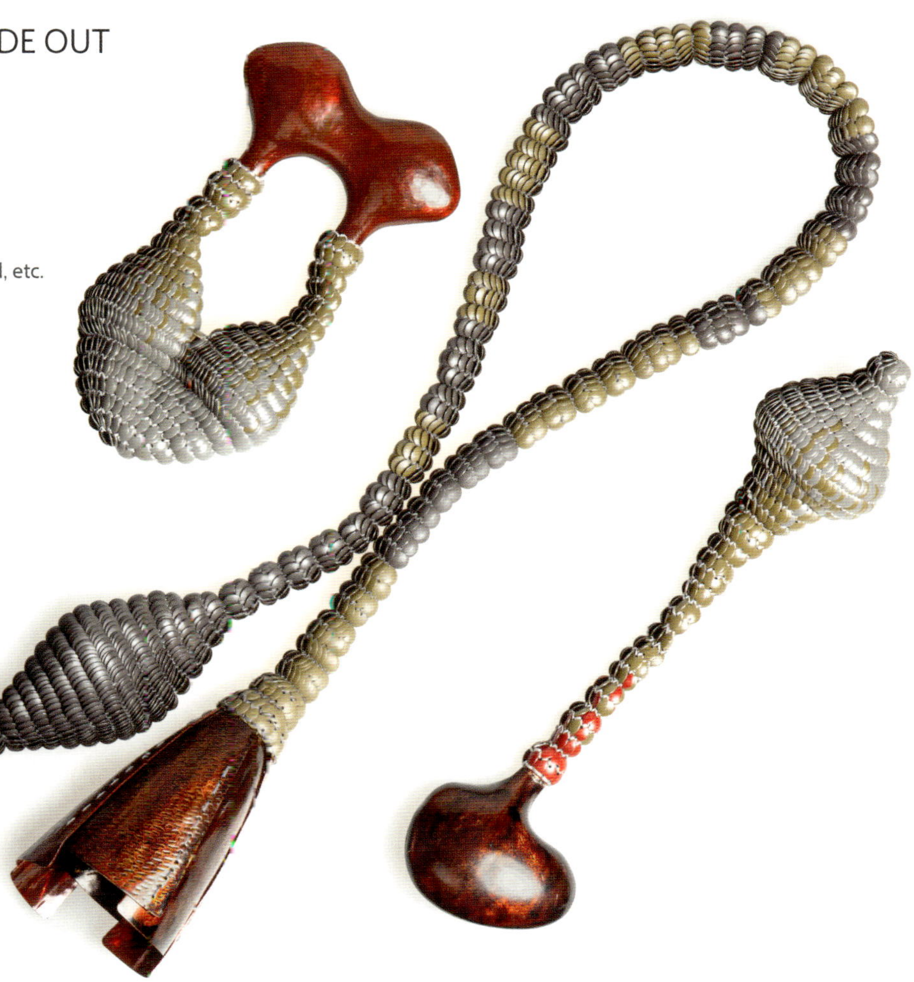

勇敢，大胆系列
BOLD, DARING SERIES

作者姓名：Jin Ah Jo（韩国）
作品类型：胸针，项链
作品材质：软钢，银，钢

Artist: Jin Ah Jo（South Korea）
Type: brooch, necklace
Material: mild-steel, silver, steel

梦想 9
DREAM IX

作者姓名：Jiseo Kim（韩国）
作品类型：项链
作品材质：皮革，金箔，金，银，石头，纸等

Artist: Jiseo Kim（South Korea）
Type: necklace
Material: leather, gold-leaf, gold, silver, stone, paper, etc.

圆柱项链
CYLINDER NECKLACE

作者姓名：Jo Pudelko（英国）
作品类型：项链
作品材质：树脂石膏基材料，铁

Artist: Jo Pudelko（Britain）
Type: necklace
Material: jesmonite, iron

气泡
BUBBLES

作者姓名：Joohee Han（韩国）
作品类型：项链
作品材质：不锈钢

Artist: Joohee Han（South Korea）
Type: necklace
Material: stainless steel

灰色孢子，橙色陷阱
SPORES GRAY , TRAP ORANGE

作者姓名：Jounghye Park（韩国）
作品类型：项链
作品材质：手工染色丝绸，925 银，塑料

Artist: Jounghye Park（South Korea）
Type: necklace
Material: hand-dyed silk, sterling silver, plastic

非塑料系列
NOT PLASTIC

作者姓名：Justine Fletcher（新西兰）
作品类型：项链
作品材质：珐琅，氧化银 925，不锈钢

Artist: Justine Fletcher（New Zealand）
Type: necklace
Material: large pendant-champleve enamel
oxidiesed 925 silver, stainless steel

海德格尔的实验室 5，海德格尔的实验室 6，海德格尔的实验室 3
HEIDEGGER'S LAB 5, HEIDEGGER'S LAB 6, HEIDEGGER'S LAB 3

作者姓名：Katja Toporski（德国）
作品类型：胸针，项链
作品材质：银，聚氨酯，光学滤波器，亚克力

Artist: Katja Toporski（Germany）
Type: brooch, necklace
Material: silver, optical filters, polyurethane, acrylic

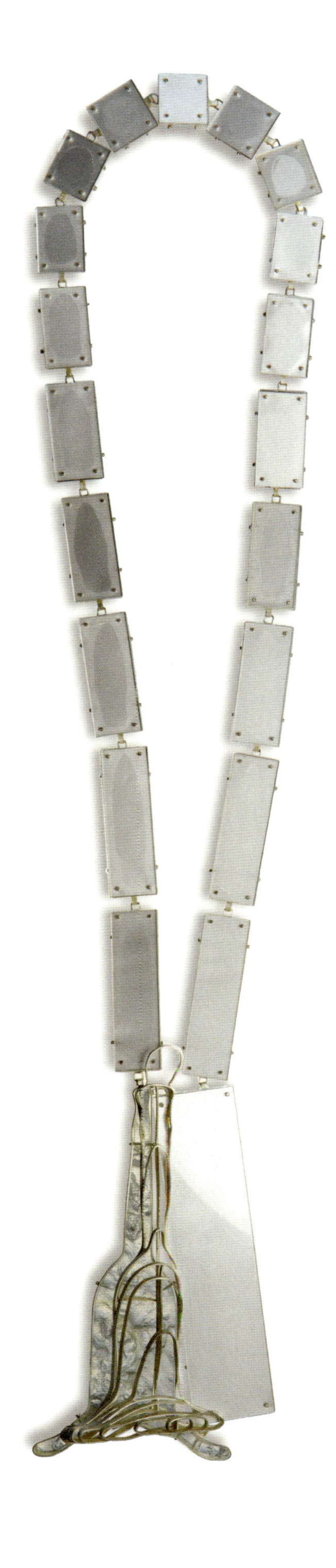

1月4日，2月4日，3月4日
JANUARY 4TH , FEBRUARY 4TH , MARCH 4TH

作者姓名：Klara Brydewall Sandquist（瑞典）
作品类型：胸针
作品材质：银，铸造核心和电铸表面

Artist: Klara Brydewall Sandquist（Sweden）
Type: brooch
Material: silver, casted core and electroformed skin

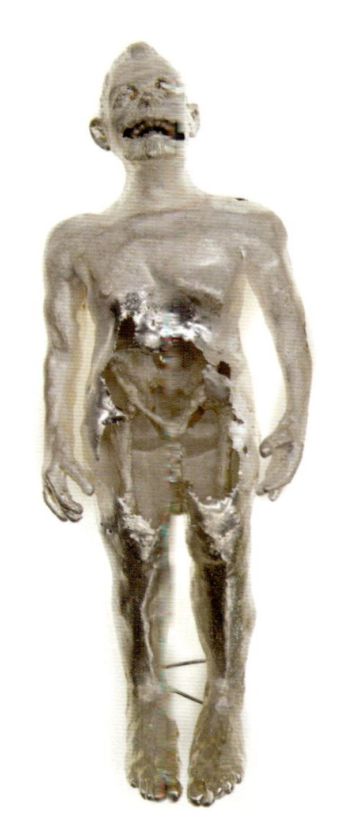

白色沉寂
SILENCE IS WHITE

作者姓名：Kristi Paap（爱沙尼亚）
作品类型：项饰，胸针
作品材质：樱桃木，温柏木，紫丁香木，缎带，银，
李子核，丹森木，油漆，绳索

Artist: Kristi Paap（Estonia）
Type: necklace, brooch
Material: cherry wood, quince wood, lilac wood, ribbon,
silver, plum stones, damson wood, paint, cord

折叠：胸针
FOLD FORM: BROOCH

作者姓名：Larah Nott（澳大利亚）
作品类型：胸针
作品材质：阳极氧化钛，不锈钢等

Artist: Larah Nott（Australia）
Type: brooch
Material: anodised titanium, stainless steel, etc.

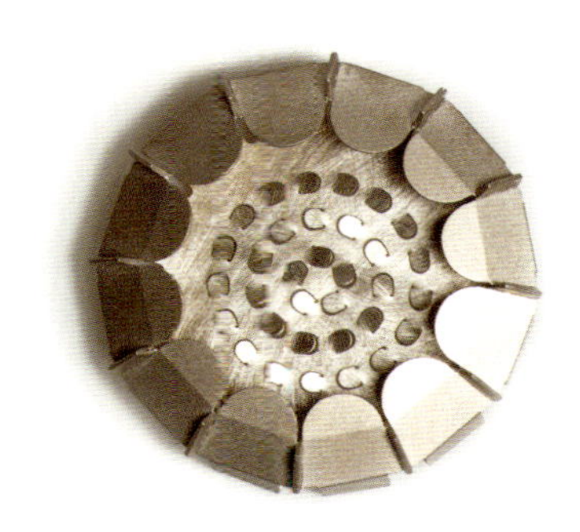

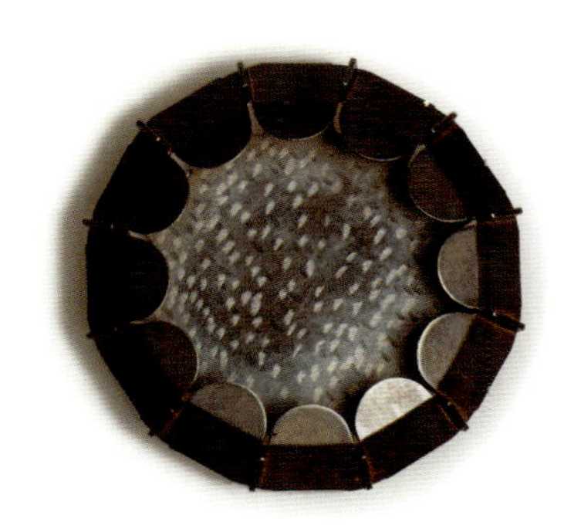

填空装置
DEVICE FOR FILLING A VOID

作者姓名：Lauren Kalman（美国）
作品类型：头饰，戒指
作品材质：镀金电铸铜

Artist: Lauren Kalman（U.S.A）
Type: headwear, ring
Material: gold-plated electroformed copper

安全食品邻里共享 1，安全食物邻里共享 2
SAFE FOOD NEIGHBOURHOOD SHARING 1， SAFE FOOD NEIGHBOURHOOD SHARING 2

作者姓名：Lavinia Rossetti（意大利）
作品类型：项链
作品材质：回收筷子，织物线，中国丝绸

Artist: Lavinia Rossetti（Italy）
Type: necklace
Material: recycled chopsticks, fabric thread, Chinese silk

碎片胸针，从 2018 年系列：修补我破碎的心
SHARDS BROOCH, 2018 FROM THE SERIES: TO MEND MY BROKEN HEART

作者姓名：Liana Pattihis（英国）
作品类型：胸针，饰针，项链
作品材质：从 6 件手工绘制的玻璃制品中选出的碎片，银链，低烧搪瓷，固色剂，不锈钢，日式杯碟瓷片，氧化银链

Artist: Liana Pattihis（Britain）
Type: brooch, necklace
Material: pieces of glass cut and randomly selected from a set of 6 oriental hand painted pieces, silver chain, low firing enamel, fixing agent, stainless steel, porcelain pieces from a Japanese cup and saucer set, oxidized silver chain

从虚无中系列
OUT OF NOTHING

作者姓名：Lital Mendel（以色列）
作品类型：戒指，胸针
作品材质：银，环氧树脂

Artist: Lital Mendel（Israel）
Type: ring, brooch
Material: silver, epoxy

深度
DEEP

作者姓名：Lucie Popelka Houdkova（捷克）
作品类型：胸针
作品材质：纸，银，不锈钢

Artist: Lucie Popelka Houdkova（Czech Republic）
Type: brooch
Material: paper, silver, stainless steel

马利亚 · 罗马纳
MAGLIA ROMANA

作者姓名：Luisa Corasaniti（意大利）
作品类型：项链
作品材质：银

Artist: Luisa Corasaniti（Italy）
Type: necklace
Material: silver

紧握，聚集
CLASPED, GATHERED

作者姓名：Lydia Martin（美国）
作品类型：胸针，项链
作品材质：925 银，18k 金等

Artist: Lydia Martin（U.S.A）
Type: brooch, necklace
Material: sterling silver, 18ct gold, etc.

自然精神系列首饰 —— 小瀑布
CASCADE NECKLACE FROM SPIRIT OF NATURE SERIES

作者姓名：Mahtab Javid（伊朗）
作品类型：胸针，项链
作品材质：梧桐树皮，纯银

Artist: Mahtab Javid (Iran)
Type: brooch, necklace
Material: bark of plane tree, fine silver

鸟生炸弹，郁金香 VII(心)，郁金香 VII(花心)
SEED BOMBS WITH BIRDS, TULIP VII (HEART), TULIP VII (FLOWER HEAD)

作者姓名：Maja Houtman（荷兰）
作品类型：胸针，项链
作品材质：925 银

Artist: Maja Houtman (Netherlands)
Type: brooch, necklace
Material: sterling silver

在这一切之后、将开始溶解，
尽管如此、它仍留在她体内，无痕

AFTER THIS EVERYTHING WILL BEGIN TO DISSOLVE，
ME WITH NO VISIBLE MARKS NEVERTHELESS
IT REMAINED INSIDE HER，IT LEFT

作者姓名：Malene Kastalje（丹麦）
作品类型：项饰，胸针
作品材质：硅树脂，颜料，钕磁铁

Artist: Malene Kastalje（Denmark）
Type: necklace, brooch
Material: silicone, pigments, neodymium magnets

特拉瓦

TRAWA

作者姓名：Maria Ignacia Walker（智利）
作品类型：项饰，胸针
作品材质：钓鱼线，树脂，银粉

Artist: Maria Ignacia Walker（Chile）
Type: necklace, brooch
Material: fishing line, resin, silver powder

长路，白色房屋，三角房屋
LONG ROAD , WHITE HOUSE , TRIANGLE HOUSE

作者姓名：Maria Rosa Franzin（意大利）
作品类型：耳环，胸针，饰针
作品材质：银，祖母绿，红珊瑚

Artist: Maria Rosa Franzin (Italy)
Type: earring, brooch
Material: silver, emeralds, red coral

转化
TRANSFORMATION

作者姓名：Mariko Sumioka（日本）
作品类型：胸针
作品材质：银，18k 金，珐琅铜，和布，竹子

Artist: Mariko Sumioka (Japan)
Type: brooch
Material: sliver,18ct gold, enamel on copper, antique kimono, bamboo

入侵者的形状
FORM OF INVADERS

作者姓名：Marina Iwagami（日本）
作品类型：胸针
作品材质：银，珐琅，色粉

Artist: Marina Iwagami (Japan)
Type: brooch
Material: silver, enamel, pastel

生长的地方
GROUND TO GROW

作者姓名：Marina Zachou（希腊）
作品类型：胸针
作品材质：青铜，尼龙打印照片，金粉

Artist:	Marina Zachou（Greece）
Type:	brooch
Material:	bronze, nailon printed photographs, gold dust

我叫弗卢斯，卡梅奥，分枝 — 无分枝
IM FLUSS CAMEO BRANCH – NO BRANCH

作者姓名：Marion Blume（德国）
作品类型：项链，胸针
作品材质：桦木胶木板，线，磁铁，橡树夹板，紫铜，钴铬合金，银枫木夹板，银，钴铬合金

Artist: Marion Blume (Germany)
Type: necklace, brooch
Material: birch veneer (wood), thread, magnet, oak veneer (wood), copper, remanium, silver maple veneer (wood), silver, remanium

永恒的戒指
FOREVER RINGS

作者姓名：Michael Jank and Bettina Dittimann（德国）
作品类型：戒指
作品材质：铁

Artist: Michael Jank and Bettina Dittimann（Germany）
Type: rings
Material: iron

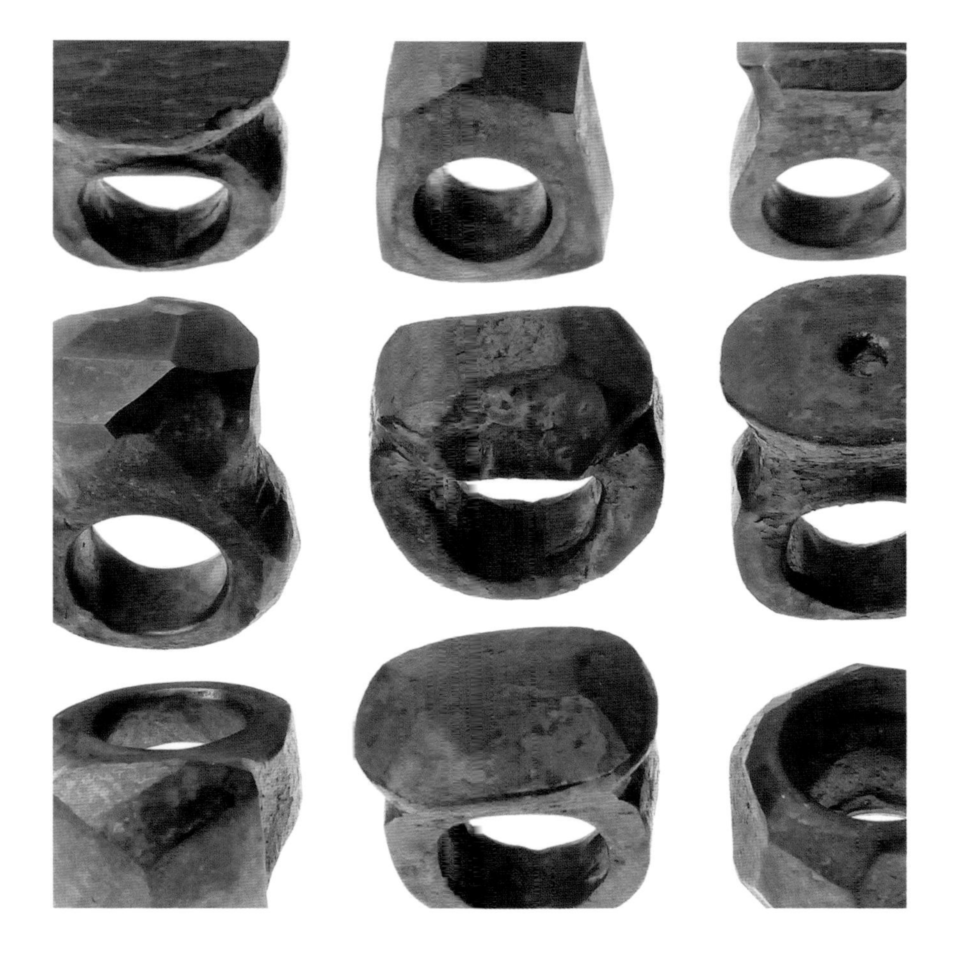

红玫瑰
THE RED ROSE

作者姓名：Michela Vincenzi（意大利）
作品类型：项链
作品材质：银，铜，丙烯漆

Artist: Michela Vincenzi（Italy）
Type: necklace
Material: silver, copper, acrilic paint

蜕变
METAMORPHOSIS

作者姓名：Mineri Matsuura（日本）
作品类型：胸针
作品材质：氧化银，不锈钢，油漆

Artist: Mineri Matsuura（Japan）
Type: brooch
Material: oxidized silver, stainless steel, paint

雨林中的飞蛾
MOTHS FROM THE RAINFOREST

作者姓名：Mio-kuhnen（澳大利亚）
作品类型：项链
作品材质：珐琅，氧化 925 银，不锈钢等

Artist: Mio-kuhnen（Australia）
Type: necklace
Material: enamel, oxidiesed 925 silver, stainless steel, etc.

灵魂碎片系列
SOUL SCRAPS SERIES

作者姓名：Miruna Belicovici（罗马尼亚）
作品类型：戒指
作品材质：手工纸，羊毛，黄铜螺丝

Artist: Miruna Belicovici（Romania）
Type: ring
Material: handmade paper, wool, brass screw

揭示，大转变，绕道
REVEAL，U-TURN，DETOUR

作者姓名：Motoko Furuhashi（美国）
作品类型：项链，胸针，手镯
作品材质：黄铜，道路油漆

Artist:	Motoko Furuhashi（U.S.A）
Type:	necklace, brooch, bracelet
Material:	brass, road paint

永恒
TIMELESS

作者姓名：Nao Mizutani（日本）
作品类型：胸针
作品材质：银

Artist:	Nao Mizutani（Japan）
Type:	brooch
Material:	sliver

一连串的无尽旅行
TRANSITION INFINITE JOURNEY

作者姓名：Nicole Schuster（德国）
作品类型：戒指，项链
作品材质：925 银，钻石，堇青石，黄水晶 氧化银，18k 金，染色丙烯酸塑料，天然漆

Artist: Nicole Schuster (Germany)
Type: ring, necklace
Material: sterling silver, diamond, cordierite, citrines, oxidised 925 silver , 18ct gold, dyed acrylic plastic, lacquer

捷克民俗（两件）
CZECH FOLKLORE

作者姓名：Nicole Taubinger（捷克）
作品类型：项链
作品材质：塑料垃圾

Artist: Nicole Taubinger (Czech Republic)
Type: necklace
Material: plastic waste

旧牛仔裤的新生！
NEW LIFE FOR MY OLD JEANS !

作者姓名：Paola Iglesias（阿根廷）
作品类型：项链
作品材质：再生牛仔布，查瓜尔纱（从查瓜尔植物中获得的天然纤维），染料，线，青铜

Artist: Paola Iglesias (Argentina)
Type: necklace
Material: recycled jean, chaguar yarn (natural fiber obtained from the chaguar plant), dye, thread, bronze

JDZ 碎片胸针
JDZ FRAGMENT BROOCH

作者姓名：Paul Wm Leathers（加拿大）
作品类型：胸针
作品材质：925 银与蛋面月光石和染色的 3D 打印尼龙

Artist: Paul Wm Leathers (Canada)
Type: brooch
Material: sterling silver with cabochon cut moonstone and dyed 3d printed nylon form

我可以吗？直到最后
AM I ABLE? UNTILL THE END

作者姓名：Paula Castro（葡萄牙）
作品类型：胸针，项链
作品材质：纸瓷，透明硬纱，银，纯银

Artist: Paula Castro (Portugal)
Type: brooch, necklace
Material: paper porcelain, organza, silver, fine silver

黑金
BLACK GOLD

作者姓名：Philipp Spillmann（瑞士）
作品类型：胸针
作品材质：银，钢针

Artist: Philipp Spillmann（Switzerland）
Type: brooch
Material: silver, steel needle

SHA- 青 I， SHA- 青 II， SHA- 青 III
SHA-GREEN I , SHA-GREEN II , SHA-GREEN III

作者姓名：Rachael Colley（英国）
作品类型：胸针，饰针
作品材质：橘子和香蕉皮，巴沙木，竹子，不锈钢

Artist: Rachael Colley（Britain）
Type: brooch
Material: orange and banana peel, balsa wood, bamboo, stainless steel

气泡项链，芝士蛋糕胸针
SPRINKLES NECKLACE , SPRINKLES CHEESECAKE BROOCH

作者姓名：Rebecca Wilson（英国）
作品类型：项链，胸针，耳环
作品材质：瓷，银，玫瑰石英，橡胶

Artist:	Rebecca Wilson (Britain)
Type:	necklace, brooch, earrings
Material:	porcelain, silver, rose quartz, rubber

底格里斯河
TIGRIS

作者姓名：Rho Tang（美国）
作品类型：胸针
作品材质：铌，纯银，不锈钢等

Artist: Rho Tang（U.S.A）
Type: brooch
Material: niobium, sterling silver, stainless steel, etc.

桥
THE BRIDGE

作者姓名：Riccardo Bonetto（意大利）
作品类型：胸针
作品材质：银

Artist: Riccardo Bonetto（Italy）
Type: brooch
Material: silver

七个小矮人
7 DWARFS

作者姓名：Ruta Naujalyte（立陶宛）
作品类型：项链，胸针
作品材质：缝纫线，丝绸，纺织线，施华洛世奇水晶

Artist: Ruta Naujalyte（Lithuania）
Type: necklace, brooch
Material: sewing threads, silk ,darning thread , swarovski crystal

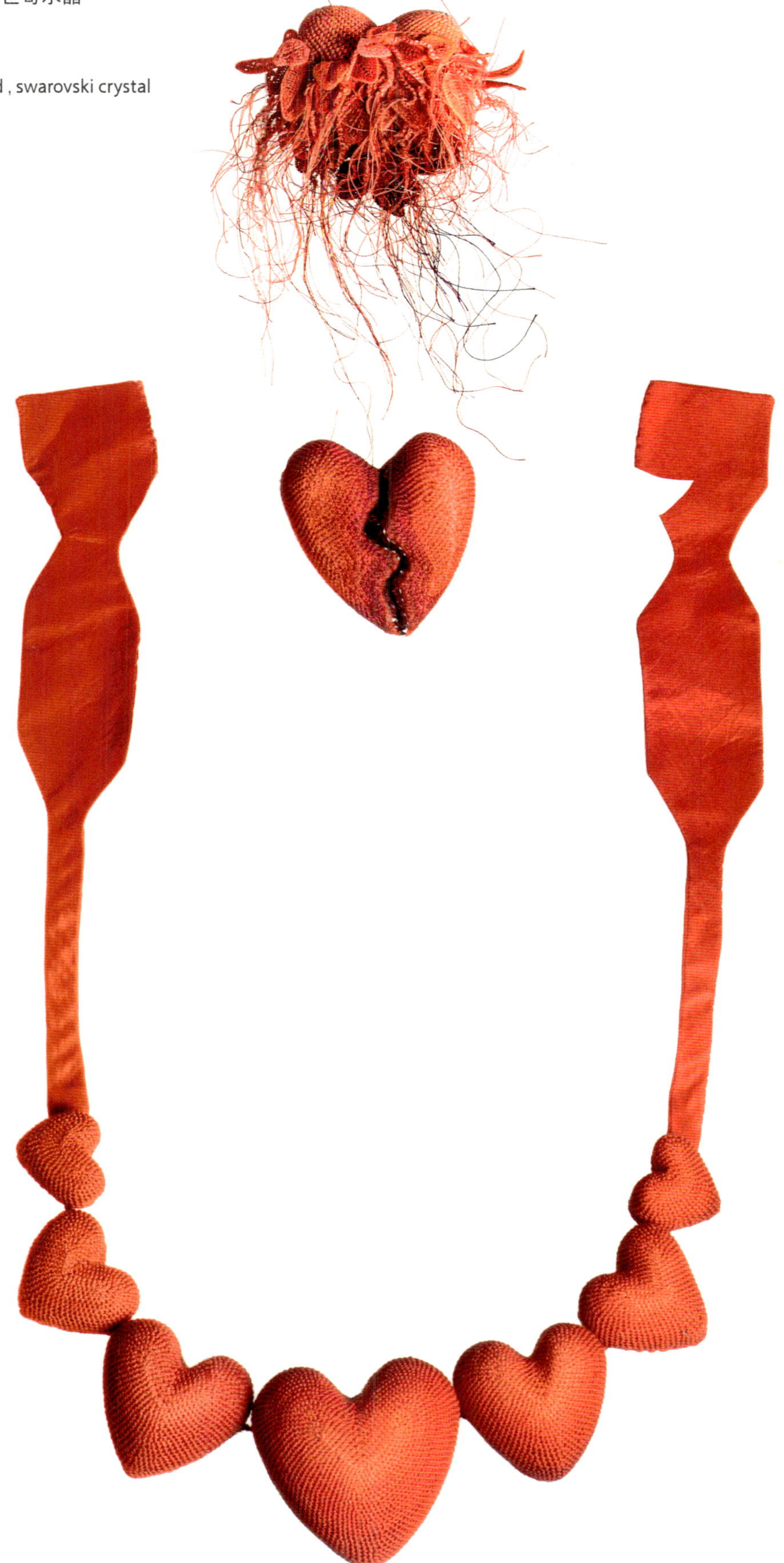

花环，无题，樱花 1
THE WREATH , UNTITLE , CHERRY BLOSSOMS 1

作者姓名：Saerom Kong（韩国）
作品类型：项链，胸针
作品材质：大米，氧化银，树脂，颜料，木头

Artist: Saerom Kong（South Korea）
Type: necklace, brooch
Material: rice, oxidized silver, resin, pigment, wood

巨变
REVOLUTION

作者姓名：Sandra Alvarado（智利）
作品类型：手镯
作品材质：氧化铜，金箔，指甲油

Artist: Sandra Alvarado (Chile)
Type: bracelet
Material: oxidized copper, gold leaf, nail polish

Resterà——给乔伊利费门托 2018
Resterà FOR GIOIELLI IN FERMENTO 2018

作者姓名：Sara Barbanti（意大利）
作品类型：胸针
作品材质：梨炭，银

Artist: Sara Barbanti（Italy）
Type: brooch
Material: pear charcoal, silver

尖锐的本性——秋季，
尖锐的本性——夏季，
尖锐的本性——冬季

SHARP NATURE-AUTUMU，
SHARP NATURE-SUMMER，
SHARP NATURE-WINTER

作者姓名：Sara Shahak-Bio（以色列）
作品类型：项链，胸针
作品材质：铁，黄铜，搪瓷，玻璃涂料，橡胶漆类，不锈钢，微型玻璃球，天然漆

Artist: Sara Shahak-Bio（Israel）
Type: necklace, brooch
Material: iron, brass, enamel, glass paint, rubber paint, stainless steel, miniature glass balls, laquere

初级生物，文化冲突，生态系统
PRIMAL ORGANISM , CULTURAL CLASH , ECOSYSTEM

作者姓名：Sébastien Carré(法国)
作品类型：手镯，胸针
作品材质：天然珍珠，小珠子，尼龙绳，日本纸，丝绸，赤铁矿，胶片，鸵鸟羽毛，绿玉髓，棉

Artist: Sébastien Carré (France)
Type: bracelet, brooch
Material: natural pearls, beads, nylon thread, Japanese paper, silk, hematite, filmstrip, ostrich feather, chrysoprase, cottcn

玛德琳
MADELEINE

作者姓名：Sofia Bankestrom（瑞士）
作品类型：胸针
作品材质：树脂，稻根，钢

Artist: Sofia Bankestrom（Switzerland）
Type: brooch
Material: resin, rice roots, steel

莲花重生，我的小波斯，流动是生活的方式
THE REBIRTH OF LOTUS, MY LITTLE PERSIA, FLUIDITY IS THE WAY TO LIFE

作者姓名：Sogand Nobahar（伊朗）
作品类型：手镯，胸针
作品材质：100% 的手工真丝波斯地毯，
黄铜 + 尼龙粉末烧结，
手工染色烧结尼龙粉 + 磁铁

Artist: Sogand Nobahar（Iran）
Type: bracelet, brooch
Material: 100% handmade silk persian,
carpet+brass+sintered nylon powder,
hand dyed sintered nylon powder+magnets

移动！沿着直线；移动！沿着曲线
MOVE! THE STRAIGHT LINE, MOVE! THE CURVED LINE

作者姓名：Sonia Pibernat（西班牙）
作品类型：胸针
作品材质：涂铜，珐琅粉

Artist: Sonia Pibernat（Spain）
Type: brooch
Material: painted copper, enamel powder

三（帆），特拉马里克，巨人
THREE（SAILS）, TERRAMARIQUE TI, EUGANELS

作者姓名：Stefano Rossi（意大利）
作品类型：胸针，项链
作品材质：木纹金（银 铜），做旧的朦胧银，银

Artist: Stefano Rossi（Italian）
Type: brooch, necklace
Material: mokume gane (silver, copper), patinated shibuichi, silver

观察者徽章
WATCHERS BADGES

作者姓名：Stephen Bottomley（英国）
作品类型：徽章
作品材质：钢，搪瓷和镍

Artist: Stephen Bottomley（Britain）
Type: badge
Material: steel, enamel and nickel

节日系列
FESTIVAL SERIES

作者姓名：Sun-Ae Kim（韩国）
作品类型：项链，胸针
作品材质：925 银，软木，薄板，线等

Artist: Sun-Ae Kim（South Korea）
Type: necklace, brooch
Material: sterling silver, cork, lamination, thread

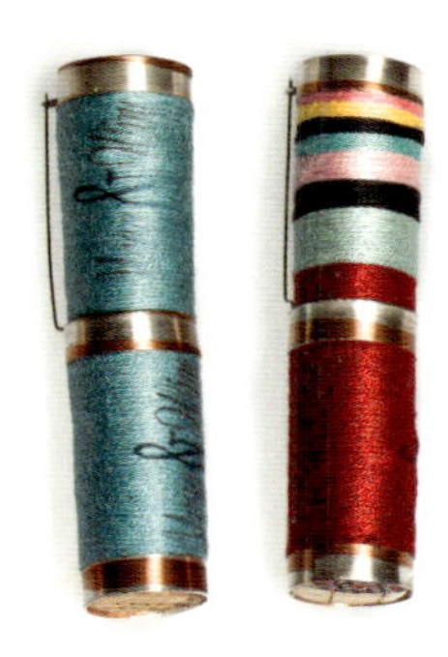

舒适的……系列
A COMFORTABLE...

作者姓名：Sun Mingrui（挪威）
作品类型：胸针
作品材质：丙烯酸分散体，回收填充物

Artist: Sun Mingrui（Norway）
Type: brooch
Material: acrylic dispersion, recycled filling

传递环
TRANSFER RING

作者姓名：Susana Teixeira（葡萄牙）
作品类型：戒指
作品材质：纯银

Artist: Susana Teixeira（Portugal）
Type: ring
Material: fine silver

几何运动，三聚氰胺，纺织连接
GEOMETRIC EXERCISES, MELAMINE, TEXTILE LINKS

作者姓名：Susanne Hammer（德国）
作品类型：项链
作品材质：木，密胺树脂，纺织品连接

Artist: Susanne Hammer（Germany）
Type: necklace
Material: wood, melamine, textile links

奥奎地亚（兰花）
ORQUIDEA (ORCHID)

作者姓名：Taibe Palacios（智利）
作品类型：胸针
作品材质：海藻，墨水，清漆，硅树脂，银，钢

Artist: Taibe Palacios（Chile）
Type: brooch
Material: seaweed, ink, varnish, silicone, silver, steel

可修复珠宝：打开空间，关闭空间
RESTORATIVE Jewellery: OPEN THE SPACE , CLOSE THE SPACE

作者姓名：Tania Cruz（墨西哥）
作品类型：戒指
作品材质：黄铜，布料

Artist: Tania Cruz（Mexico）
Type: ring
Material: copper, textile

CWAB：利恩 · 9 号
CWAB: LIEN #9

作者姓名：Teresa Faris（美国）
作品类型：胸针
作品材质：标准银，被鸟啄过的木头，回收的宠物鸟栖息杆

Artist: Teresa Faris（U.S.A）
Type: brooch
Material: sterling silver, wood altered by a bird, reclaimed comfy perch

低碳钢奖章
MILD STEEL BADGES

作者姓名：Tim Carson（英国）
作品类型：物件
作品材质：软钢，不锈钢，黄铜，粉末涂料，磁铁，钢，铬

Artist: Tim Carson（Britain）
Type: object
Material: mild steel, staintess steel, brass, powder coat, magnets, steel, chrome

为这个地球
OF THIS EARTH

作者姓名：Toni Mayner（英国）
作品类型：坠饰
作品材质：回收的铁桶带，纯金，亚麻线，纯银

Artist: Toni Mayner (Britain)
Type: pendant
Material: re-claimed iron barrel band, fine gold, linen thread, sterling silver

秘密花园
SECRET GARDEN

作者姓名：Ute Van Der Plaats（德国）
作品类型：戒指，项链
作品材质：骨瓷，纸瓷，银

Artist: Ute Van Der Plaats (Germany)
Type: ring, necklace
Material: bone china, paper porcelain, silver

面具人格，清洁完成，胸骨
PERSONA, DONE CLEANING, STERNUM

作者姓名：Veronica Cheann（挪威）
作品类型：项链，胸针
作品材质：银，纺织品，成型的领子

Artist: Veronica Cheann (Norway)
Type: necklace, brooch
Material: silver, textile, ready-made collar

棱锥道
PIRAMID WAY

作者姓名：Veronica Santello（意大利）
作品类型：项链
作品材质：银，铜，铜绿

Artist: Veronica Santello（Italy）
Type: necklace
Material: silver, copper, patina

历史分析，海螺，多元宇宙字谜
PARATIME HISTORIA , HELICA , MULTIVERSE REBUS

作者姓名：Viktoria Münzker（斯洛伐克）
作品类型：胸针，项链
作品材质：浮木，海胆骨架，珍珠

Artist: Viktoria Münzker（Slovakia）
Type: brooch, necklace
Material: driftwood, sea urchin skeleton, pearl

经典——一套六枚胸针，经典系列 1
CLASSIC—SET OF SIX BROOCHES , CLASSIC #1

作者姓名：Yael Friedman（以色列）
作品类型：胸针
作品材质：橙色开孔橡胶，不锈钢

Artist: Yael Friedman（Israel）
Type: brooch
Material: orange open cell rubber, stainless stell

缠绕，生长，脉管
WIND , SPRING , VESSEL

作者姓名：Yasmin Vinograd（以色列）
作品类型：项链
作品材质：中国瓷器，银，黑陶器

Artist: Yasmin Vinograd（Israel）
Type: necklace
Material: Chinese porcelaine, silver, basalt

实体，终端
ESSE , RTU

作者姓名：Yoko Takirai（日本）
作品类型：手镯，项链
作品材质：925 银，不锈钢

Artist:	Yoko Takirai（Japan）
Type:	bracelet, necklace
Material:	sterling silver, stainless steel

发光的身体系列
LIGHT EMITTING BODY SERIES

作者姓名：Yu Hiraishi（日本）
作品类型：胸针
作品材质：黄铜，聚氨酯漆，不锈钢

Artist: Yu Hiraishi（Japan）
Type: brooch
Material: brass，urethane paint，stainless steel

WORKS OF

CHINESE ARTISTS

中国参展艺术家作品

北京国际首饰艺术展 2019

东西星球
EAST AND WEST PLANET

作者姓名：NK 东西宫旗下设计师（中国）
作品类型：手镯，耳环，戒指
作品材质：星光蓝宝石，蓝宝石

Artist: Designer Of Nk East West Palace（China）
Type: bracelet, earring, ring
Material: star sapphire, sapphire

空间，当怪物来敲门，生长
SPACE，WHEN THE MONSTER KNOCKS，GROWING

作者姓名：车嘉妍（中国）
作品类型：胸针
作品材质：925银，PC镜，毛毡

Artist: Che Jiayan（China）
Type: brcoch
Material: sterling silver, PC mirror, flet

言系列
YAN SERIES

作者姓名：陈彬雨（中国）
作品类型：胸针
作品材质：银，锆石，纤维材料，乌木，珍珠

Artist: Chen Binyu（China）
Type: brooch
Material: silver, zircon, fiber material, ebony, pearl

苗韵凝方
MIAO YUN NING FANG

作者姓名：陈承洁（中国）
作品类型：项饰，手镯，耳饰
作品材质：银

Artist: Chen Chengjie (China)
Type: necklace, bracelet, earrings
Material: silver

穹顶系列
DOME SERIES

作者姓名：陈海宇（中国）
作品类型：胸针
作品材质：银，铜

Artist: Chen Haiyu (China)
Type: brooch
Material: silver, copper

曲律
RHYTHM OF THE SONG

作者姓名：陈嘉慧（中国）
作品类型：项饰
作品材质：925 银，陶瓷，珍珠

Artist: Chen Jiahui（China）
Type: necklace
Material: sterling silver, ceramic, pearl

万方安和之一
WANFANG ANHE

作者姓名：陈敏（中国）
作品类型：项饰
作品材质：银

Artist: Chen Min（China）
Type: necklace
Material: silver

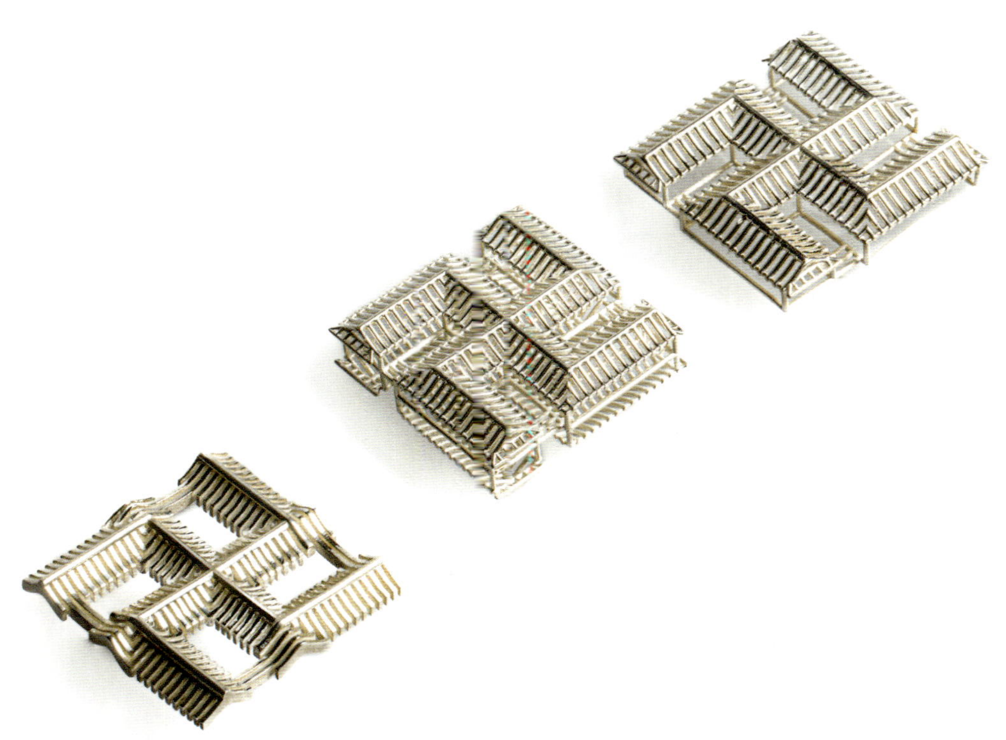

缠绕的线
TWIN THREAD

作者姓名：陈尚仪（中国）
作品类型：项饰，戒指
作品材质：银，锆石，棉线

Artist: Chen Shangyi (China)
Type: nacklace, ring
Material: silver, zircon , cotton thread

暗中窥视
MONITORING IN THE DARK

作者姓名：陈书铭（中国）
作品类型：头饰
作品材质：乳胶布，偏光镜片，银镀金

Artist: Chen Shuming (China)
Type: headpiece
Material: latex cloth, polarized lens, gold-plated silver

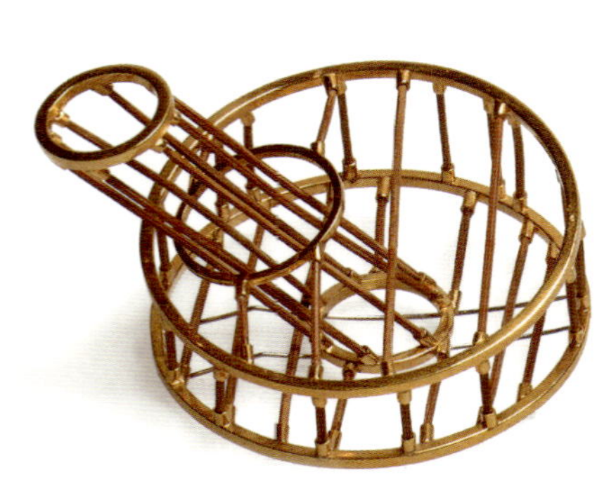

人造轨道 05， 人造轨道 06 ， 冬日的颂歌
ARTIFICIAL ORBIT 05 , ARTIFICIAL ORBIT 06 , ODE TO WINTER

作者姓名：陈素雨（中国）
作品类型：项链，胸针
作品材质：PVC 管，925 银，丙烯，酸漆

Artist: Chen Suyu（China）
Type: necklace, brooch
Material: pvc tube, sterling silver, stainless steel, acrylic paint

话筒
MICROPHONE

作者姓名：陈伟丽（中国）
作品类型：项饰
作品材质：999 银，925 银，澳大利亚 100% 纯羊毛

Artist: Chen Weili（China）
Type: necklace
Material: fire silver, sterling silver,
100% pure wool in Australia

鹦系列
PARROT SERIES

作者姓名：陈欣（中国）
作品类型：项饰
作品材质：银，紫铜等

Artist: Chen Xin（China）
Type: necklace
Material: silver, red copper, etc.

生
BORN

作者姓名：陈芷雅（中国）
作品类型：项饰
作品材质：玻璃，黄铜镀金

Artist: Chen Zhiya (China)
Type: necklace
Material: glass, gold-plated brass

平湖秋月，山前梅花

AUTUMN MOON OVER THE PEACEFUL LAKE, PLUM BLOSSOM IN FRONT OF MOUNTAIN

作者姓名：程园（中国）
作品类型：戒指
作品材质：18k 金，翡翠，黄钻，钻石

Artist: Cheng Yuan (China)
Type: ring
Material: 18ct gold, jade, yellow diamond, diamond

身体虚拟域
THE VIRTUAL FIELD

作品姓名：程之璐（中国）
作品类型：时尚配饰
作品材质：3D 打印 PLA 树脂，铜

Artist: Cheng Zhilu（China）
Type: fashion jewellery
Material: 3D printing PLA resin, brass

白描系列——之一
TRADITIONAL CHINESE LINE-DRAWING

作品姓名：崔金玉（中国）
作品类型：胸针
作品材质：综合材料

Artist: Cui Jinyu（China）
Type: brooch
Material: mixed material

鸿蒙系列吊坠（七件套之一）
HONGMENG SERIES PENDANT (ONE OF SEVEN SETS)

作者姓名：丁晓飞（中国）
作品类型：吊坠
作品材质：铜胎珐琅

Artist: Ding Xiaofei（China）
Type: pendant
Material: copper, enamel

看着的，她
LOOKING AT HER

作者姓名：丁雪妍（中国）
作品类型：项饰
作品材质：银

Artist: Ding Xueyan（China）
Type: necklace
Material: silver

隙
CRACK

作品姓名：段丙文（中国）
作品类型：项饰
作品材质：银，珐琅

Artist: Duan Bingwen（China）
Type: necklace
Material: silver, enamel

调料瓶 1
CONDIMENT BOTTLES OF 1

作者姓名：段永慧（中国）
作品类型：戒指
作品材质：紫铜，胡桃木，镀银

Artist: Duan Yonghui（China）
Type: ring
Material: copper, walnut, silver plating

7 × 7：时间内向局外人
7 × 7:TIME INSIDE STRANGER

作者姓名：方淳加（中国）
作品类型：胸针
作品材质：杜邦纸，丝线，银

Artist: Fang Chunjia（China）
Type: brooch
Material: dupont paper, silk thread, silver

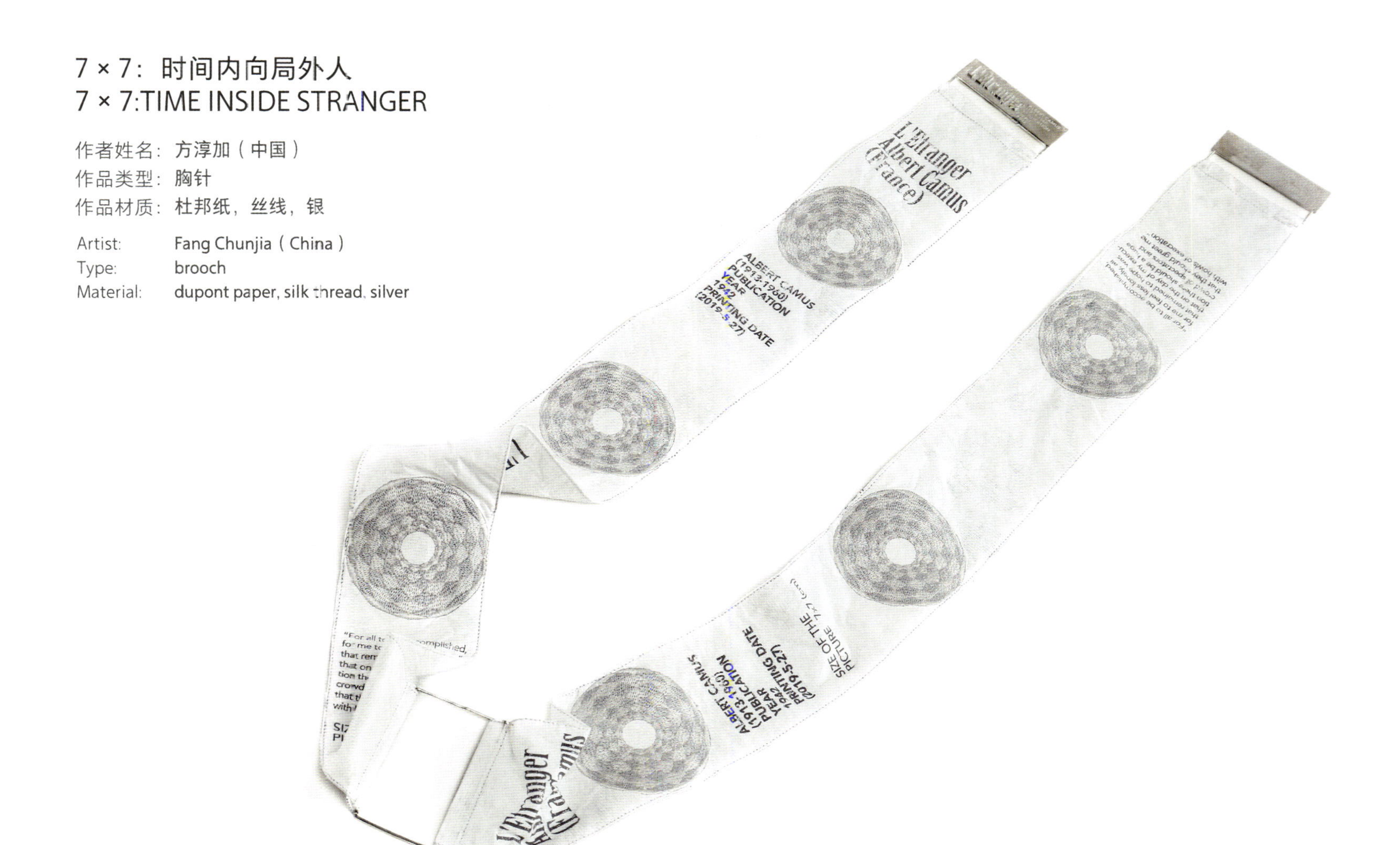

咸酸之外
BEYOND SALTY ACID

作者姓名：方龙慧子（中国）
作品类型：胸针
作品材质：银，猫毛，鱼线，亚克力，树脂等

Artist: Fang Longhuizi（China）
Type: brooch
Material: silver, cat hair, fishing line, acrylic, resin, etc.

我
ME

作者姓名：方笑晗（中国）
作品类型：其他
作品材质：玳瑁，铜，镀金

Artist: Fang Xiaohan（China）
Type: other
Material: tortoise shell, copper, gold-plated

呼吸系列
BREATHE SERIES

作者姓名：费博（中国）
作品类型：项饰
作品材质：天然漆，螺钿，蛋壳，金粉，银箔

Artist: Fei Bo（China）
Type: necklace
Material: lacquer, mother-of-pearl inlay, eggshell, gold powder, silver foil

那些花儿，国色
THE FLOWERS，NATIONAL BEAUTY

作者姓名：冯雪晶（中国）
作品类型：胸针，项饰
作品材质：银，珐琅，翡翠，珍珠

Artist:	Feng Xuejing（China）
Type:	brooch, necklace
Material:	silver, enamel, jadeite, pearl

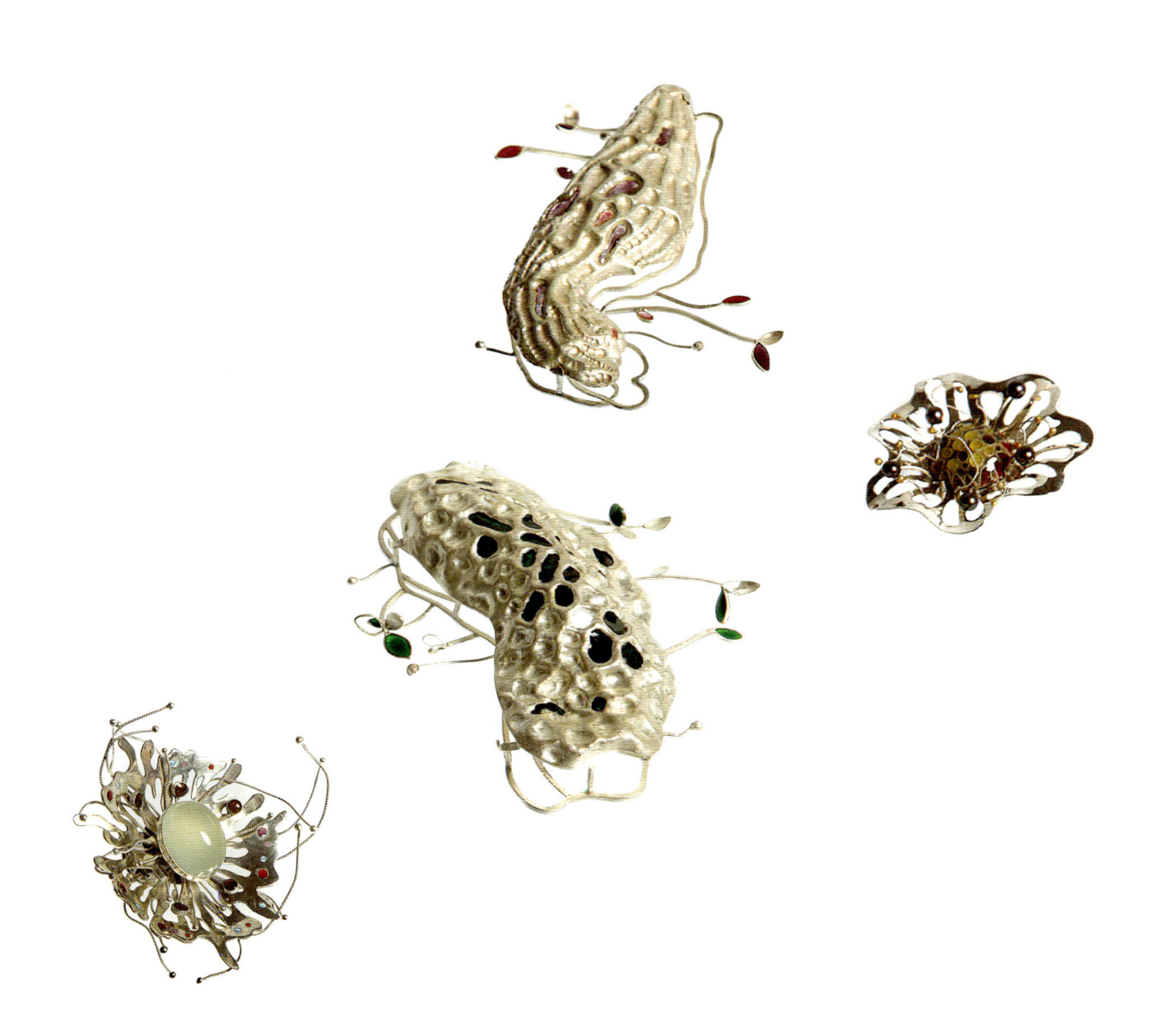

宁可食无肉
BETTER TO EAT WITHOUT MEAT

作者姓名：傅永和（中国）
作品类型：项链
作品材质：紫铜，925 银，珐琅

Artist: Fu Yonghe（China）
Type: necklace
Material: red copper，sterling silver，enamel

动之以画——桃鸠图
MOVE TO DRAW – PEACH DOVE DIAGRAM

作者姓名：傅渝卓（中国）
作品类型：耳饰
作品材质：黄铜镀金

Artist: Fu Yuzhuo (China)
Type: earrings
Material: gold-plated brass

聚光
SPOTLIGHT

作者姓名：高珊（中国）
作品类型：胸针
作品材质：999 银，黄铜，金箔，树脂，二级灯管

Artist: Gao Shan (China)
Type: brooch
Material: 999 silver, brass, gold foil, resin, two stage lamp tube

大风系列
BLUSTERY SERIES

作者姓名：高伟（中国）
作品类型：胸针
作品材质：白玉，金，银

Artist: Gao Wei (China)
Type: brooch
Material: white jade, gold, silver

舞者 —— 魔术师戒指，舞者 —— 魔术手镯
DANCER-MAGICAN RINGS, DANCER-MAGICAN BANGLE

作者姓名：高艺霖（中国）
作品类型：戒指，手镯
作品材质：不锈钢镀金，黄铜镀金，水晶，黑曜石，蛋白石

Artist:	Gao Yilin（China）
Type:	ring, bracelet
Material:	gold-plated stainless steel,gold-plated brass, crystal, obsidian, opal

一叶舟，一尘举
A TINY BOAT, A RAISING DUST

作者姓名：宫平（中国）
作品类型：胸针，手镯
作品材质：18k 金，玉石，银，石，综合材料

Artist: Gong Ping（China）
Type: brooch, bracelet
Material: 18ct gold, jade, silver, stone, mixed material

垃圾计划
GARBAGE PLAN

作者姓名：巩志伟（中国）
作品类型：胸针
作品材质：废旧办公用纸，紫铜，黄铜，925 银

Artist: Gong Zhiwei (China)
Type: brooch
Material: waste office paper, red copper, brass, sterling silver

新生
NEW LIFE

作者姓名：古丽米拉 · 艾尼（中国）
作品类型：胸针
作品材质：925 银，彩色锆石

Artist: Gulimila · Aini (China)
Type: brooch
Material: sterling silver, colored zircon

可以不可以
YES OR NO

作者姓名：谷明（中国）
作品类型：胸针
作品材质：钛金属，椰壳，和田玉

Artist: Gu Ming（China）
Type: brooch
Material: titanium, coconut husk, nephrite

童年与职业
CHILDHOOD AND OCCUPATION

作者姓名：郭靖凯（中国）
作品类型：戒指
作品材质：925 银，珍珠

Artist: Guo Jingkai（China）
Type: ring
Material: sterling silver, pearl

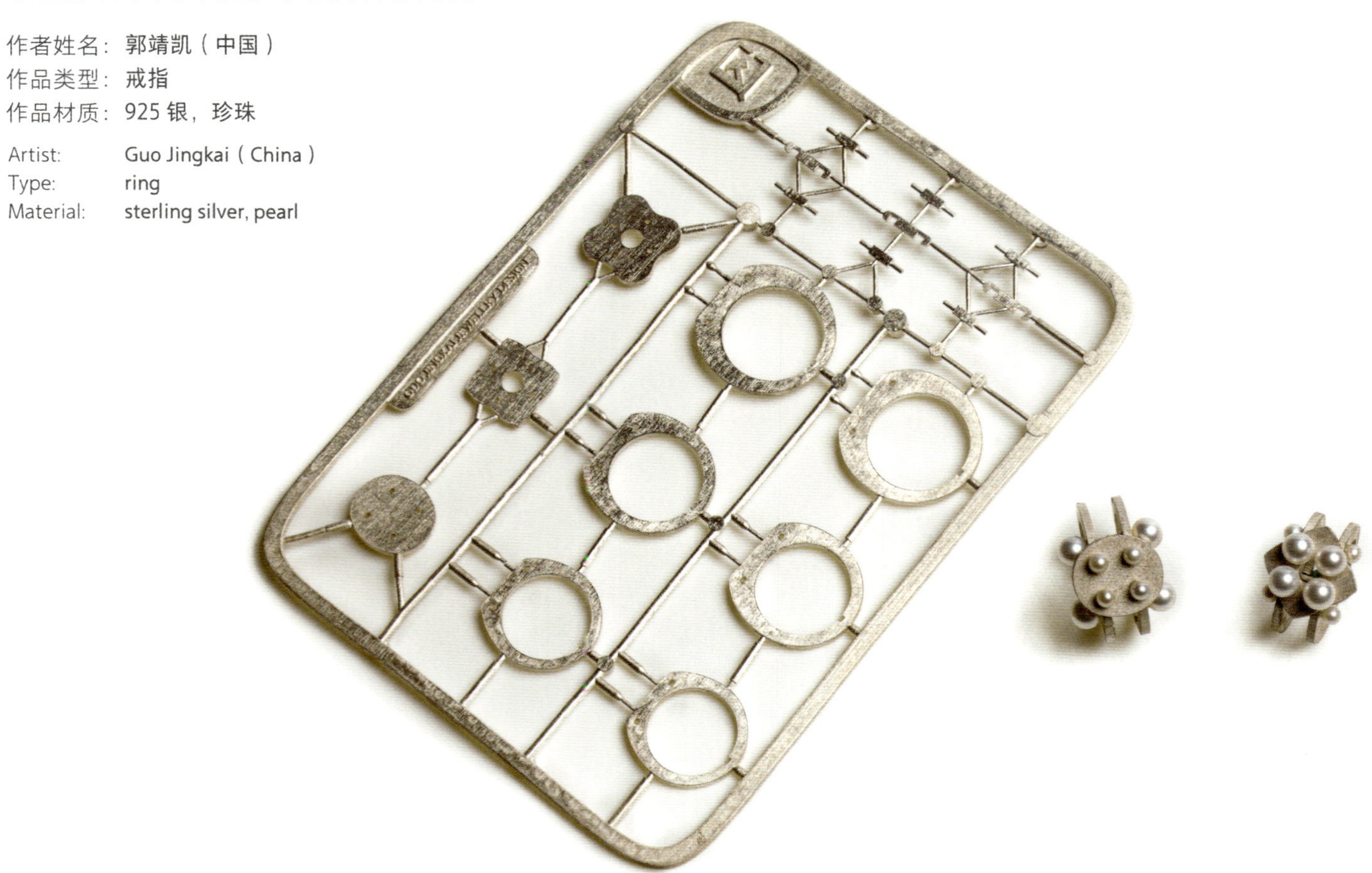

PM 2.5

作者姓名：郭强（中国）
作品类型：物件
作品材质：925 银

Artist: Guo Qiang（China）
Type: object
Material: sterling silver

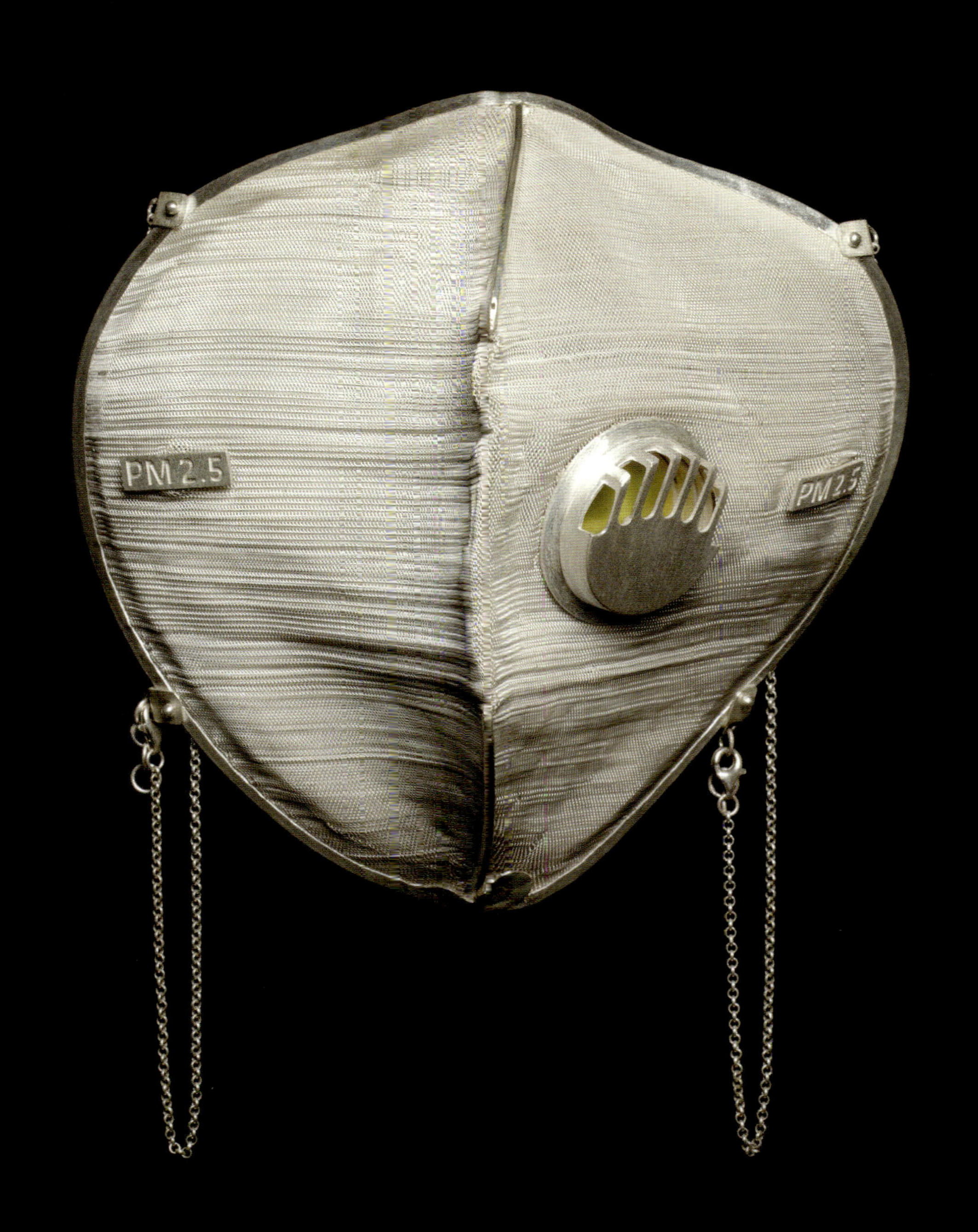

蜕变
METAMORPHOSIS

作者姓名：郭晓枫（中国）
作品类型：手镯
作品材质：电铸铜胎冷珐琅，水晶滴胶

Artist: Guo Xiaofeng (China)
Type: bracelet
Material: electrocast copper tire cold enamel, crystal epoxy

蓝色的回响
BLUE ECHO

作者姓名：郭之箐（中国）
作品类型：项饰
作品材质：树脂，丝线

Artist: Guo Zhiqing (China)
Type: necklace
Material: resin, silk thread

24 小时粉碎迷恋
24HRS OF SMASHED OBSESSION

作者姓名：郭芷欣（中国）
作品类型：耳饰
作品材质：925 银

Artist: Guo Zhixing（China）
Type: earrings
Material: sterling silver

萨满巫师——狼牙
SHAMAN WIZARD -WOLF TOOTH

作者姓名：韩冬（中国）
作品类型：项饰
作品材质：骨骼，乌木，青铜箭头

Artist: Han Dong (China)
Type: necklace
Material: bone, ebony, bronze arrowhead

獵巫行动头饰
WITCH HUNT HEADGEAR

作者姓名：韩乐遥（中国）
作品类型：头饰
作品材质：黄铜，雪纺绸

Artist: Han Leyao (China)
Type: headwear
Material: brass, chiffon

相由心生
THE FACE IS THE INDEX OF THE HEART

作品姓名：韩欣然（中国）
作品类型：胸针
作品材质：光敏树脂，铜，金箔

Artist:	Han Xinran（China）
Type:	brooch
Material:	photosensitive resin , copper, gold foil

木卫系列
JUPITER SERIES

作者姓名：韩雨蒙（中国）
作品类型：戒指
作品材质：航空铝，银，树脂，珐琅，珍珠

Artist: Han Yumeng（China）
Type: ring
Material: aviation aluminum, silver, resin, enamel, pearl

香港的镜头系列
LENS-SCAPE OF HONG KONG SERIES

作者姓名：何霭英（中国香港）
作品类型：戒指
作品材质：火山岩，黄铜凸透镜，石板

Artist:	Ying Valerie (Hongkong, China)
Type:	ring
Material:	volcanic rock, brass and convex lens, slate

共体，春暖花开
THE TOTAL BODY , SPRING FLOWERS

作者姓名：胡俊（中国）
作品类型：胸针
作品材质：925 银，金箔，钢丝，天然漆，树脂

Artist: Hu Jun（China）
Type: brooch
Material: sterling silver, gold foil, steel wire, lacquer, resin

无题
UNTITLED

作者姓名：胡世法（中国）
作品类型：胸针，项饰
作品材质：银，钢，珊瑚，珍珠

Artist: Hu Shifa（China）
Type: brooch, necklace
Material: silver, steel, coral, pearl

视己台
SEE YOURSELF

作者姓名：胡钰鋆（中国）
作品类型：胸针
作品材质：黄铜，925 银，亚克力

Artist: Hu Yuyun（China）
Type: brooch
Material: brass, sterling silver, acrylic

收集梦境—1
DREAM COLLECTION-1

作者姓名：黄秋韵（中国）
作品类型：其他
作品材质：黄铜镀，24k 金，珍珠，红色锆石，指甲油，925 银

Artist: Huang Qiuyun（China）
Type: other
Material: brass coating, 24ct gold, pearl, red zircon, nail polish, sterling silver

买一送一（免费的那一个只是倒影）
BUY ONE GET ONE FREE (THE FREE ONE IS IN THE REFLECTION)

作者姓名：黄逸（中国）
作品类型：耳饰
作品材质：铜镀金，珍珠

Artist: Huang Yi（China）
Type: earrings
Material: gold-plated copper, pearl

参与我的在线广告，H，U，交织
ENGAGE WITH MY ONLINE AD, H , U, INTERWEAVE

作者姓名：简毓奇（中国台湾）
作品类型：戒指，耳饰，手镯
作品材质：纯银，黄金，伦敦蓝田玉等

Artist: Yuchi Chien（Taiwan, China）
Type: ring, earring, bracelet
Material: fine silver, gold, London blue topaz, etc.

绽放，共生体
BLOOMING , SYMBIONT

作者姓名：姜倩（中国）
作品类型：项饰，胸针，戒指
作品材质：银，木头，木皮，毛毡

Artist: Jiang Qian（China）
Type: necklace, brooch, ring
Material: silver, wood, veneer, felt

转换
TRANSFORMATION

作者姓名：蒋悦（中国）
作品类型：项链，戒指，耳饰
作品材质：925 银，珍珠，塑料，苔藓，棉

Artist: Jiang Yue (China)
Type: necklace, ring, earring
Material: sterling silver, pearl, plastic, moss, cotton

律动
RHYTHM

作者姓名：金翠玲（中国）
作品类型：胸针
作品材质：925 银，纸

Artist: Jin Cuiling (China)
Type: brooch
Material: sterling silver, paper

丝路花雨
THE SILK ROAD

作者姓名：晋文捷（中国）
作品类型：项饰
作品材质：黄金，925 银，滴胶，珐琅彩，丙烯

Artist: Jin Wenjie (China)
Type: necklace
Material: gold, sterling silver, epoxy, enamel, propylene

梦呓系列之一
ONE OF THE DREAMILY SERIES

作者姓名：乐钊（中国）
作品类型：胸针
作品材质：银

Artist: Yue Zhao (China)
Type: brooch
Material: silver

21世纪玉组佩
JADE GROUP IN 21ST CENTURY

作者姓名：李安琪（中国）
作品类型：项饰
作品材质：软玉，玉线

Artist: Li Anqi（China）
Type: necklace
Material: nephrite, jade thread

帆
THE SAIL

作者姓名：李登登（中国）
作品类型：胸针
作品材质：银

Artist: Li Dengdeng（China）
Type: brooch
Material: silver

爱是纯真的童年
LOVE IS INNOCENT CHILDHOOD

作者姓名：李恒（中国）
作品类型：项饰
作品材质：白铜镀白金，绣线，汽车烤漆，玻璃细珠

Artist: Li Heng（China）
Type: necklace
Material: platinum plating on white copper, embroidery thread, paint, glass beads

小表情系列
SMALL EXPRESSION SERIES

作者姓名：李菁（中国）
作品类型：胸针
作品材质：银

Artist: Li Jing（China）
Type: brooch
Material: silver

黑洞
THE BLACK HOLE

作者姓名：李静（中国）
作品类型：戒指
作品材质：铜电镀黑金

Artist: Li Jing (China)
Type: ring
Material: black gold-plated copper

双生子
TWINS

作者姓名：李莉（中国香港）
作品类型：项饰
作品材质：钛金属，白钻，115 克拉无烧蓝宝，翡翠，刚玉

Artist: Li Li (Hong Kong, China)
Type: necklace
Material: titanium, white diamond, 115 carats blue gemstones, jade, corundum

叶
LEAF

作者姓名：李鹏（中国）
作品类型：项饰
作品材质：银，碧玺

Artist: Li Peng（China）
Type: necklace
Material: silver, tourmaline

让我进洞，行走的珍珠，扭曲的珍珠
GET ME INTO THE HOLE , WALKING PEARLS , TWISTING PEARL

作者姓名：李嫱（中国）
作品类型：戒指
作品材质：镀金黄铜，天然巴托克珍珠，磁石，锆土，木，丙烯，不锈钢

Artist: Li Qiang（China）
Type: ring
Material: gold-plated on brass, natural baroque pearls, magnets, zircon, wood, acrylic, stainless steel

亲爱的鹿
WHITE PEAK DEER DEAR!

作者姓名：李然（中国）
作品类型：胸针
作品材质：树脂，颜料，银

Artist: Li Ran（China）
Type: brooch
Material: resin, pigment, silver

路
ROAD

作者姓名：李桑（中国）
作品类型：项饰
作品材质：银

Artist: Li Sang（China）
Type: necklace
Material: silver

寻求
SEEK

作者姓名：李天清（中国）
作品类型：胸针
作品材质：铜，银，植物

Artist: Li Tianqing (China)
Type: brooch
Material: copper, silver, plants

残存之物
REMNANTS

作者姓名：李小筠（中国）
作品类型：胸针
作品材质：天然漆，银，螺钿

Artist: Li Xiaoyun（China）
Type: brooch
Material: lacquer, silver, mother of pearl

默契交流
TACTILE COMMUNICATION

作者姓名：李亦佳（中国）
作品类型：项链
作品材质：树脂石膏基材料，绳线

Artist: Sissel Li（China）
Type: necklace
Material: jesmonite, thread

海错罐头系列
CANNED SEA CREATURES

作者姓名：李颖臻（中国）
作品类型：胸针，项饰
作品材质：珐琅，生牛皮，银，麻

Artist: Li Yingzhen (China)
Type: brooch, necklace
Material: enamel, rawhide, silver, linen

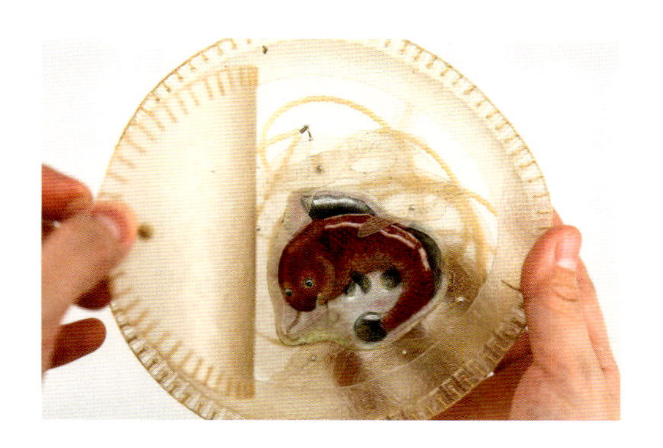

黑洞
BLACK HOLE

作者姓名：李昀倩（中国）
作品类型：胸针
作品材质：光固化树脂

Artist: Li Yunqian (China)
Type: brooch
Material: light cured resin

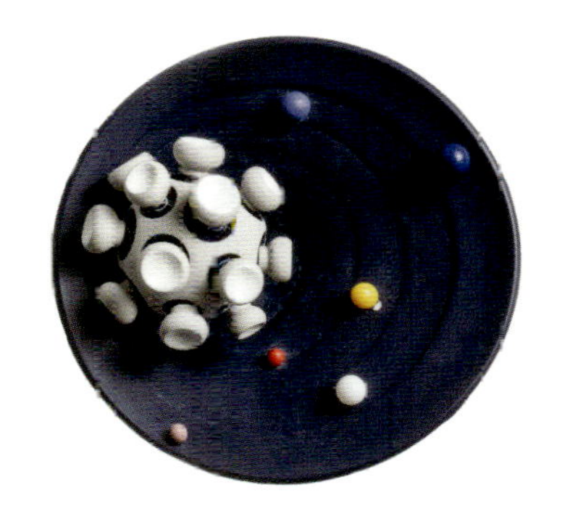

爱，浪漫，热烈
LOVE , ROMANTIC, FERVENT

作者姓名：李哲（中国）
作品类型：胸针
作品材质：银，纸

Artist: Li Zhe (China)
Type: brooch
Material: sliver, paper

烦恼
WORRY

作者姓名：李卓怡（中国）
作品类型：戒指
作品材质：银，米，塑料，铅笔

Artist: Li Zhuoyi (China)
Type: ring
Material: silver, rice, plastic, pencil

权力的创守，傲世华姿，蓄势高飞
THE ESTABLISHMENT OF POWER, AO SHI HUA ZI, READY TO FLY HIGH

作者姓名：林弘裕（中国）
作品类型：吊坠，胸针
作品材质：18k 白金，方形翡翠，钻石

Artist: Lin Hongyu（China）
Type: pendant, brooch
Material: 18ct platinum, square jade, diamond

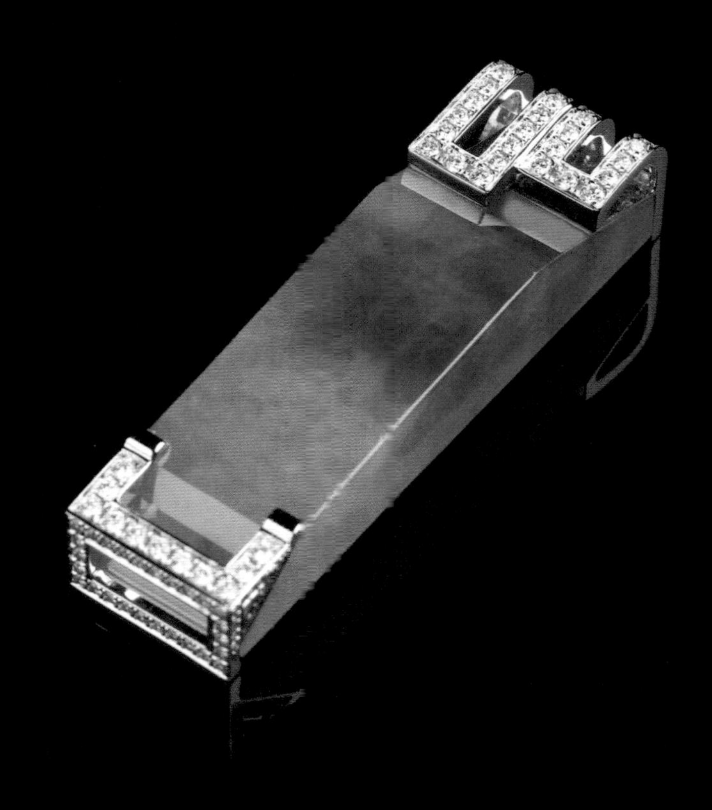

锦鲤
LUCKY CYPRINOID

作者姓名：刘过（中国）
作品类型：胸针
作品材质：925 银，18k 金，24k 金等

Artist: Liu Guo (China)
Type: brooch
Material: sterling silver, 18ct gold, 24ct gold, etc.

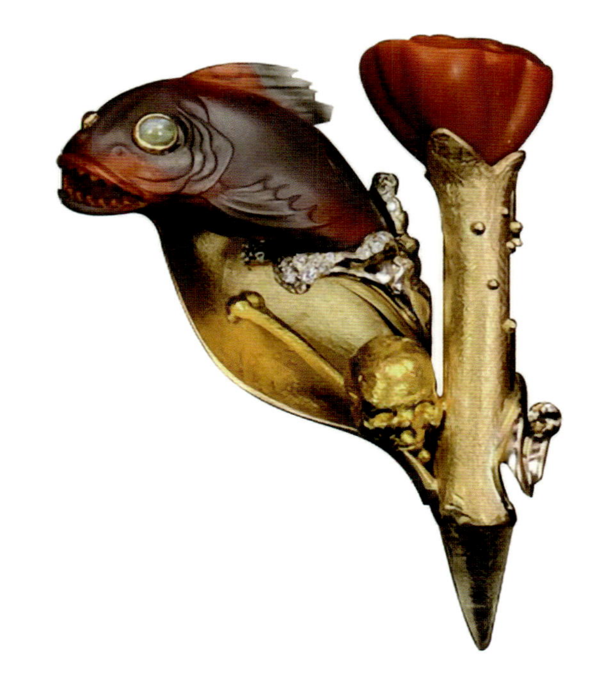

集，森林精灵，相承
COLLECTION, WOOD ELVES, CARRY ON

作者姓名：刘娇（中国）
作品类型：项链
作品材质：银，珍珠，木材，珊瑚，玻璃，树脂

Artist: Liu Jiao (China)
Type: necklace
Material: silver, pearl, wood, coral, glass, resin

时光
TIME

作者姓名：刘静（中国）
作品类型：耳坠
作品材质：925 银，铜，钛

Artist: Liu Jin（China）
Type: earrings
Material: sterling silver, copper, titanium

界限之间
ON EDGE

作者姓名：刘小奇（中国）
作品类型：项饰，耳饰，胸针
作品材质：合金，925 银

Artist: Liu Xiaoqi（China）
Type: necklace, earring, brooch
Material: alloy, sterling silver

游园惊梦飞鸟翡翠项链
SWEET DREAM IN THE GARDEN FLYING BIRD NECKLACE

作者姓名：龙梓嘉（中国）
作品类型：项链（吊坠可做胸针用）
作品材质：翡翠，蓝宝石，玉石，青金石等

Artist: Long Zijia（China）
Type: necklace（pendent can be used as brooch）
Material: jadeite, sapphire, jade, green treasure, etc.

冬天的童话
THE FAIRY TALE IN THE WINTER

作者姓名：卢艺（中国）
作品类型：胸针
作品材质：钛，纱，珍珠，绒花，碳纤维

Artist: Lu Yi（China)
Type: brooch
Material: titanium, yarn, pearl, velvet followers, carbon fiber

寻觅
SEEK

作者姓名：鲁硕（中国）
作品类型：摆件
作品材质：紫水晶，925 银

Artist: Lu Shuo（China）
Type: decoration
Material: amethyst, sterling silver

云银深处
DEEP IN CLOUD AND SILVER

作者姓名：罗理亭（中国）
作品类型：胸针
作品材质：紫光檀，银，金

Artist: Luo Liting（China）
Type: brooch
Material: African sandalwood, silver, gold

项链——纳贡贮贝器，项链——鹰，胸针——虎
A CONTAINER FOR PAYING TRIBUTE, NECKLACE-THE EAGLE, A BROOCH-TIGER

作者姓名：罗元园（中国）
作品类型：项饰，胸针
作品材质：金，银，树脂

Artist: Luo Yuanyuan（China）
Type: necklace, brooch
Material: gold, silver, resin

无题
UNTITLED

作者姓名：孟海晨（中国）
作品类型：臂钏
作品材质：银，皮革

Artist: Meng Haichen（China）
Type: armlet
Material: silver, leather

死灰复燃
THE RESURGENCE

作者姓名：潘杨（中国）
作品类型：戒指
作品材质：综合材料，烧焦的木头，黄铜，蜡

Artist: Pan Yang（China）
Type: ring
Material: mixed material, burnt wood, brass, wax

纯之爱
PURE LOVE

作者姓名：裴潇雨（中国）
作品类型：耳环
作品材质：黄钻

Artist: Pei Xiaoyu（China）
Type: earrings
Material: yellow diamond

悬着的心，太阳还是那么美，父爱如山
HANGING HEART, THE SUN STILL BEAUTIFUL, FATHER'S LOVE IS AS GREAT AS MOUNTAINS

作者姓名：屈梦楠（中国）
作品类型：胸针
作品材质：24k 金，天然漆，925 银，999 银

Artist: Qu Mengnan（China）
Type: brooch
Material: 24ct gold, lacquer, sterling silver, fine silver

花间住
LIVING IN THE FLOWERS

作者姓名：任俊颖（中国）
作品类型：胸针
作品材质：铜，银，胧银，赤铜

Artist: Ren Junying (China)
Type: brooch
Material: copper, silver, oboro silver, red copper

长河系列，幻影系列
CHANGHE SERIES, PHANTOM SERIES

作者姓名：深圳市盛峰黄金有限公司（中国）
作品类型：手镯
作品材质：极光超硬金，足金

Artist: Shenzhen Shengfeng Gold Co., Ltd.（China）
Type: bracelet
Material: aurora superhard gold, full gold

希望在 6 月17日
HOPE ON 17TH JUNE

作者姓名：时俊（中国）
作品类型：胸针
作品材质：纸黏土，宣纸，墨，金箔

Artist: Shi Jun（China）
Type: brooch
Material: paper clay, xuan paper, ink, gold foil

午夜的病
THE MIDNIGHT DISEASE

作者姓名：史晨超（中国）
作品类型：项饰
作品材质：玻璃纤维，记忆软胶，银

Artist: Shi Chenchao（China）
Type: necklace
Material: glass fiber, shape memory soft glue, silver

嗨！ 听！ #1#2#3
HEY! LISTEN! #1#2#3

作者姓名：史湘吟（中国）
作品类型：胸针
作品材质：925 银，找到的物品，树脂，丙烯酸塑料，钢

Artist: Shi Xiangyin（China）
Type: brooch
Material: sterling silver, found objects, resin, acrylic paint, steel

那山系列
THE MOUNTAIN SERIES

作者姓名：史忠文（中国）
作品类型：胸针
作品材质：925 银

Artist: Shi Zhongwen（China）
Type: brooch
Material: sterling silver

记忆里的糖
SUGAR IN THE MEMORY

作者姓名：帅思澄（中国）
作品类型：胸针
作品材质：银，树脂

Artist: Shuai Sicheng (China)
Type: brooch
Material: silver, resin

观云系列
WATCHING CLOUDS SERIES

作者姓名：孙谷藏（中国）
作品类型：项饰，胸针
作品材质：银，陶瓷

Artist: Sun Guzang (China)
Type: necklace, brooch
Material: silver, ceramic

龙秘，睚眦
DRAGON MYTH, YA ZI

作者姓名：孙浩洋（中国）
作品类型：胸针
作品材质：银，沉香木

Artist: Herman Sun（China）
Type: brooch
Material: silver, agilawood

万花筒
KALEIDOSCOPE

作者姓名：孙静茹（中国）
作品类型：项饰
作品材质：银黄铜，鲍鱼贝

Artist: Sun Jingru（China）
Type: necklace
Material: brass, abalone shell

晟·姬，婠·剡
SHENG · JI, WAN · YAN

作者姓名：孙平（中国）
作品类型：项饰，胸针
作品材质：银，珍珠，玻璃

Artist: Sun Ping（China）
Type: necklace, brooch
Material: silver, pearl, glass

城市狐狸村，城市狐狸森林，城市狐狸城
CITY FOX VILLAGE , CITY FOX FOREST , CITY FOX CITY

作者姓名：孙以诺（中国）
作品类型：戒指
作品材质：银

Artist: Sun Yinuo（China）
Type: ring
Material: silver

一个女孩和她自己，
一个男孩和她自己，蓝色
A GIRL AND HERSELF ,
A BOY AND HERSELF , BLUE

作者姓名：谭瑶（中国）
作品类型：项链
作品材质：陶瓷，珍珠，铁，黄铜，蕾丝缎带，玻璃珠

Artist: Tan Yao (China)
Type: necklace
Material: ceramic, pearl, iron, brass, bud silk ribbon, glass bead

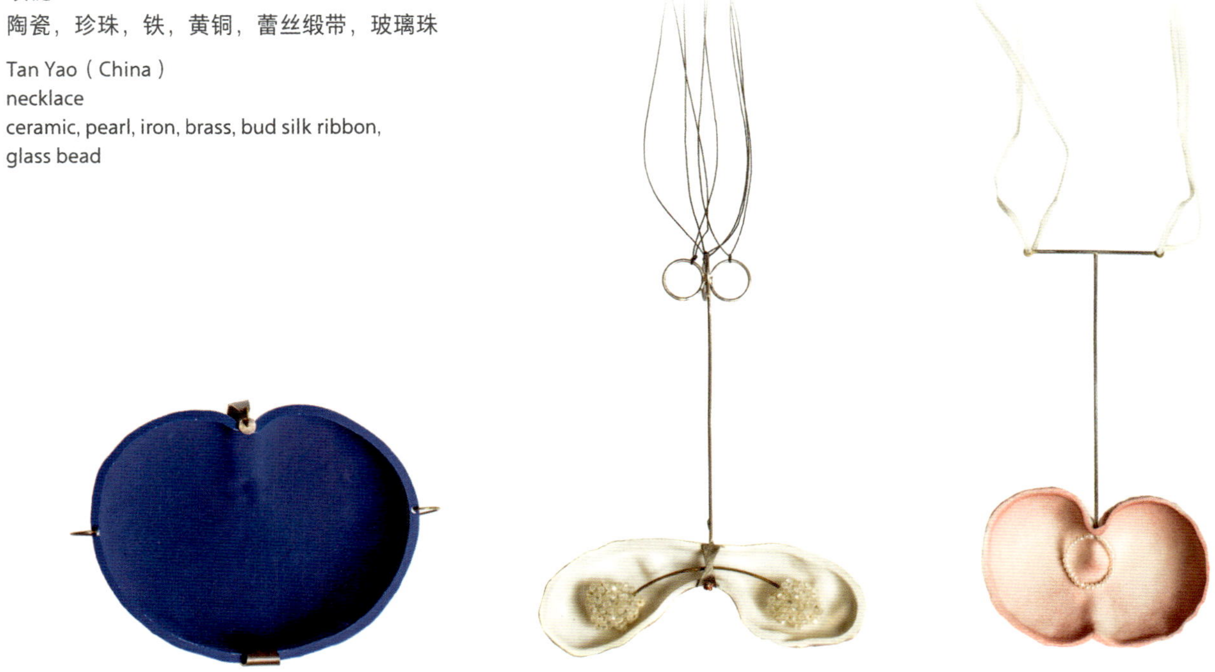

铜鼓行——古韵声生
BRONZE DRUM LINE-ANCIENT RHYME

作者姓名：唐超，马磊（中国）
作品类型：戒指，手镯，其他
作品材质：银，皮

Artist: Tang Chao , Ma Lei (China)
Type: ring, bracelet, other
Material: silver, leather

无题
UNTITLED

作品姓名：汪正红（中国）
作品类型：项链
作品材质：特种纸，黄水晶，925 银，18k 镀金

Artist: Wang Zhenghong（China）
Type: necklace
Material: special paper, topaz, sterling silver, 18ct gold-plated

啮
GNAW

作者姓名：王春刚（中国）
作品类型：项饰
作品材质：银，珐琅，珍珠

Artist: Wang Chungang（China）
Type: necklace
Material: silver, enamel, pearl

金镶玉——飘逸的蝴蝶兰
GOLD COVERS JADE FLOWING PHALAENOPSIS

作者姓名：王海涛（中国）
作品类型：胸针
作品材质：和田玉，18k 金，红宝石

Artist: Wang Haitao（China）
Type: brooch
Material: nephrite, 18ct gold, ruby

图像失真
IMAGE DISTORTION

作者姓名：王浩睿（中国）
作品类型：奖牌
作品材质：纯银

Artist: Wang Haorui（China）
Type: medal
Material: fine silver

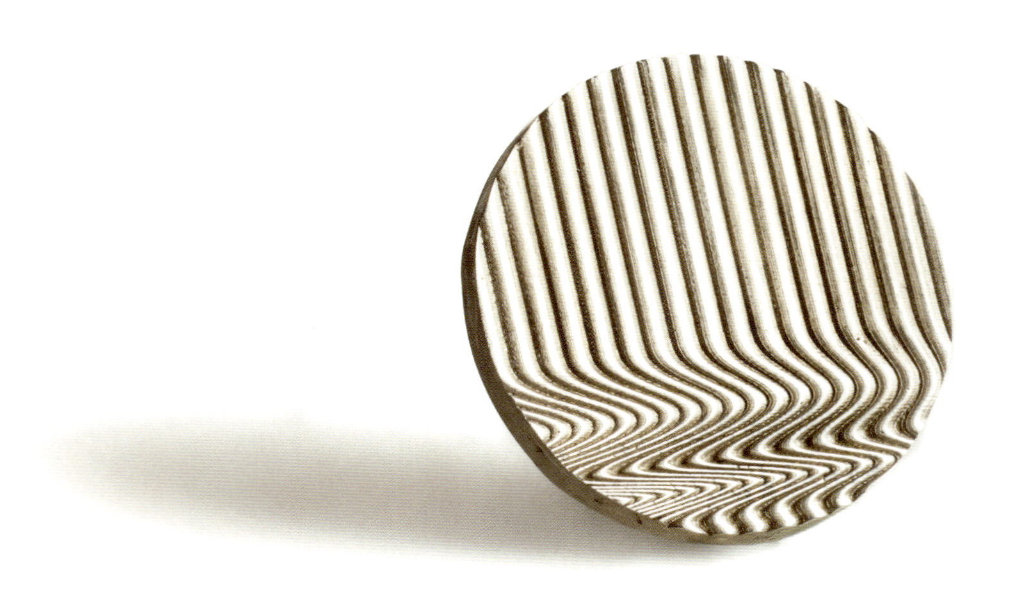

夫载福履系列
FUZAIFULU SERIES

作者姓名：王嘉暄（中国）
作品类型：胸针
作品材质：银

Artist: Wang Jiaxuan（China）
Type: brooch
Material: silver

春华秋实，镜花如幻
GLORIOUS FLOWERS IN SPRIING AND SOLID FRUITS IN AUTUMN， FLOWERS IN THE MIRROR ARE LIKE FANTASY

作者姓名：王敬（中国）
作品类型：胸针
作品材质：银，综合材料，钛金属着色，珐琅

Artist: Wang Jing（China）
Type: brooch
Material: silver, mixed materials, titanium plating, enamel

脸书
FACEBOOK

作者姓名：王克震（中国）
作品类型：胸针
作品材质：骆驼骨，925 银

Artist: Wang Kezhen（China）
Type: brooch
Material: camel bone, sterling silver

永恒（从肚脐到手中）
ETERNITY (FROM NAVEL TO HAND)

作者姓名：王玲婕（中国）
作品类型：戒指
作品材质：铜，铜镀银，铜镀金

Artist: Wang Lingjie（China）
Type: ring
Material: copper, silver plated copper, gold-plated copper

海上生明月耳环
MOONRISE BY THE SEA

作者姓名：王圣临（中国）
作品类型：耳环，胸针吊坠两用
作品材质：18k 白金，akya 珍珠，钻石

Artist: Wang Shenglin（China）
Type: earrings, brooch pendant dual-purpose
Material: 18ct platinum, akya pearl, diamond

钻石与陶瓷
DIAMOND AND CERAMIC

作者姓名：王泰迪（中国）
作品类型：胸针
作品材质：18k 金，老瓷片

Artist:	Wang Taidi（China）
Type:	brooch
Material:	18ct gold, old tiles

沙与沫
SAND AND FOAM

作者姓名：王涛（中国）
作品类型：胸针
作品材质：白铜

Artist: Wang Tao（China）
Type: brooch
Material: cupronickel

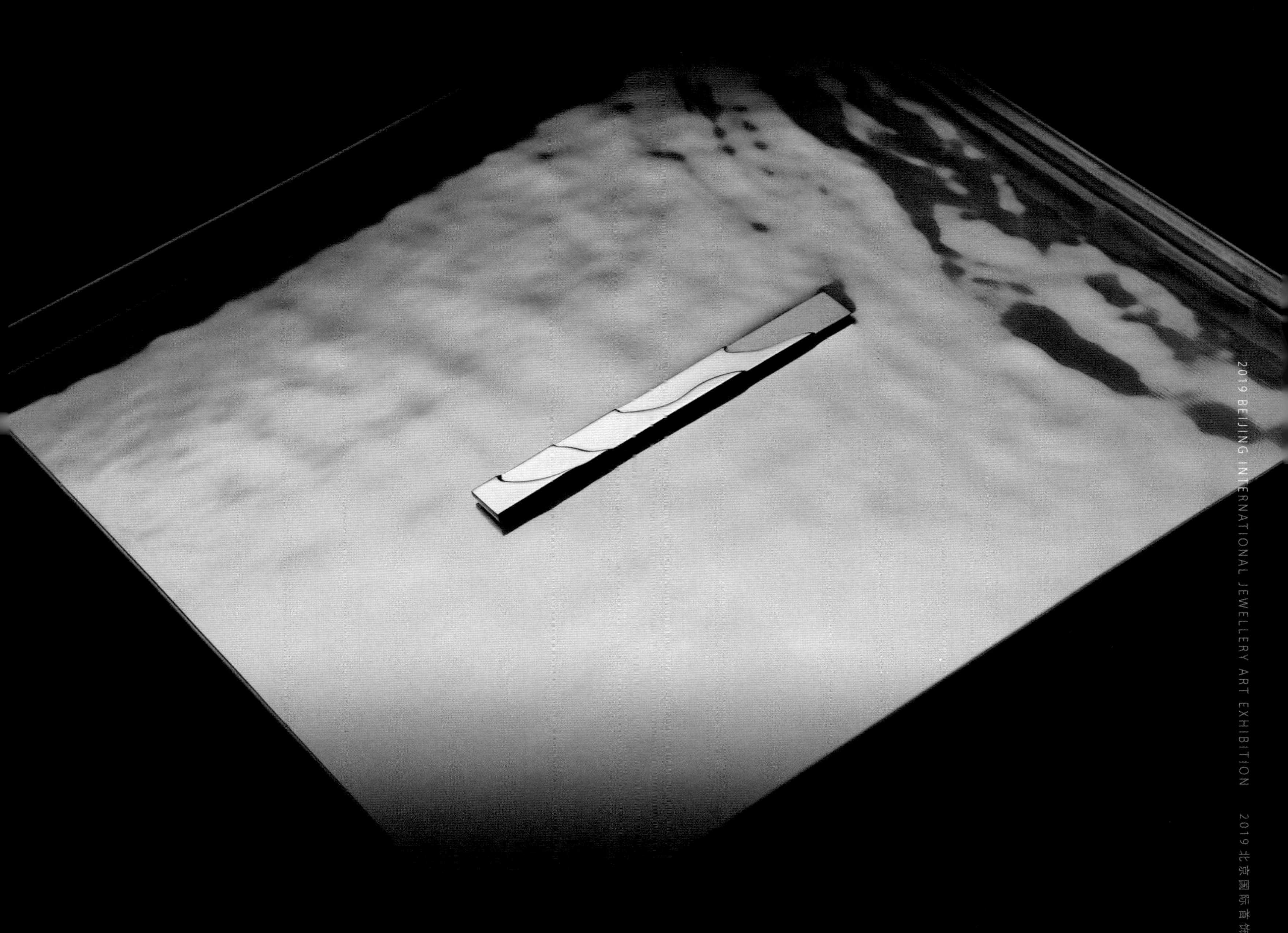

清光疏影
PURE LIGHT WITH SPARSE SHADOW

作者姓名：王晓昕（中国）
作品类型：胸针
作品材质：混凝土，黄铜，金

Artist: Wang Xiaoxin（China）
Type: brooch
Material: concrete, brass, gold

生命中不能承受之轻
UNBEARABLE LIGHTNESS IN LIFE

作者姓名：王笑佳（中国）
作品类型：胸针
作品材质：宣纸，墨，木头，羽毛，漆等

Artist: Wang Xiaojia (China)
Type: brooch
Material: xuan paper, ink, wood, feather, paint, etc.

城市密码
CITY PASSWORD

作者姓名：王莹（中国）
作品类型：项饰，手镯，戒指
作品材质：银，玛瑙

Artist: Wang Ying (China)
Type: necklace, bracelet, ring
Material: silver, agate

二仔
THE SECOND SON

作者姓名：王泽丹（中国）
作品类型：吊坠
作品材质：18k 玫瑰金，钻

Artist: Wang Zedan（China）
Type: pendant
Material: 18ct rose gold, diamond

华特尔，石头二号
WARATAH , ROCKSII

作者姓名：韦思腾（中国）
作品类型：胸针
作品材质：环氧树脂，925 银，不锈钢，透明胶片，聚丙烯

Artist: Siteng Wei（China）
Type: brooch
Material: epoxy resin, sterling silver, stainless steel, transparency film，polypropylene

插入一个石头心脏，行人 2
PLUG IN A STONE HEART , PASSERBY 2

作者姓名：吴冬怡（中国）
作品类型：胸针
作品材质：黏土，铅笔，开关水龙头，插头，925 银，不锈钢，布，铜，皮革，丙烯颜料，塑料

Artist: Wu Dongyi (China)
Type: brooch
Material: clay, pencil, switch tap, plug, sterling silver, stainlees steel, fabic, copper, leather, acrylic pigment, plastic

黑色的巢
THE BLACK NEST

作者姓名：吴芳（中国）
作品类型：项饰
作品材质：软陶，纱，珍珠，数据线

Artist: Wu Fang（China）
Type: necklace
Material: polymer clay, gauze, pearl, USB cable

丰收
HARVEST

作者姓名：吴捍（中国）
作品类型：项饰
作品材质：18k 金，钻石，碧玺

Artist: Wu Han（China）
Type: necklace
Material: 18ct gold, diamond, tourmaline

复生
RESURRECTION

作者姓名：吴冕（中国）
作品类型：戒指，胸针
作品材质：银，金箔

Artist: Wu Mian（China）
Type: brooch, ring
Material: silver, gold foil

行道树系列，城市痕迹
STREET TREES SERIES , CITY TRACE

作者姓名：吴树玉（中国）
作品类型：胸针，耳饰
作品材质：铜，珐琅，925 银，钢针等

Artist: Wu Shuyu（China）
Nationality: brooch, earrings
Type: copper, enamel, sterling silver,
Material: steel needle, etc.

憧憬
LONGING

作者姓名：伍艺麒（中国）
作品类型：项链
作品材质：吉他片拨片，18k 金

Artist: Wu Yiqi（China）
Type: necklace
Material: guitar picks, 18ct gold

陌上花开系列
MO SHANG HUA KAI SERIES

作者姓名：肖尧（中国）
作品类型：项饰，手镯，戒指
作品材质：纯银，18k 铜镀金，人工宝石

Artist: Xiao Yao（China）
Type: necklace, bracelet, ring
Material: fine silver, 18ct gold-plated copper, artificial gem

共生
COMMENSALISM

作者姓名：谢馥蔚（中国）
作品类型：戒指
作品材质：银，锆石

Artist: Xie Fuwei（China）
Type: ring
Material: sliver, zircon

世界的边缘 1
THE EDGE OF THE WORLD I

作者姓名：谢隽（中国）
作品类型：项链
作品材质：人工合成材料，铜，塔拉纳基，铁砂

Artist: Xie Jun（China）
Type: necklace
Material: synthetic material, copper, taranaki, iron sand

汉蝶秋风
BATTERFLY OF HAN AND WIND OF AUTUMN

作者姓名：熊芏芏（中国）
作品类型：戒指
作品材质：木头，钛，银，铜

Artist: Xiong Dudu（China）
Type: ring
Material: wood, titanium, silver, copper

交错，旋转
STAGGERED , ROTATE

作者姓名：徐可（中国）
作品类型：戒指
作品材质：18k 金，石榴石，彩色蓝宝石等

Artist: Xu Ke（China）
Type: ring
Material: 18ct gold, garnet, coloured sapphire, etc.

断片系列
FRAGMENT SERIES

作者姓名：徐玫莹（中国）
作品类型：项饰，胸针
作品材质：竹，竹炭，纸，丝线，黄铜等

Artist: Xu Meiying（China）
Type: necklace, brooch
Material: bamboo, bamboo charcoal, paper, silk thread, brass, etc.

山雪未霁
THE MOUNTAIN SNOW HAS NOT YET LIFTED

作者姓名：徐倩（中国）
作品类型：项链
作品材质：银，铜，珐琅

Artist: Xu Qian（China）
Type: necklace
Material: silver, copper, enamel

言系列
YAN SERIES

作者姓名：许安然（中国）
作品类型：胸针
作品材质：银，墨玉

Artist: Xu Anran（China）
Type: brooch
Material: silver, black jade

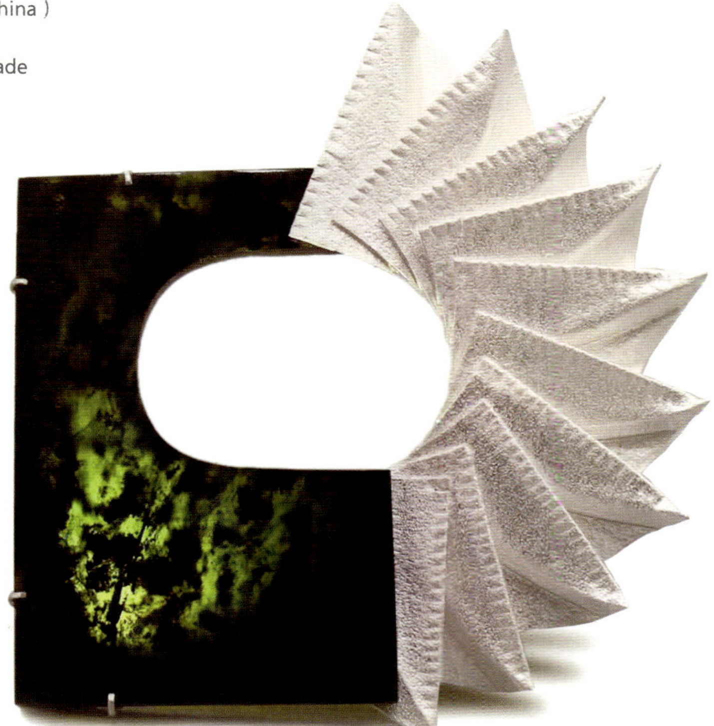

都市美型男系列
METROSEXUAL SERIES

作者姓名：许国蕤（中国）
作品类型：戒指，手镯，项饰
作品材质：925 银电镀黑金，黄金，白金，马尾毛，立方氧化锆

Artist: Xu Guorui（China）
Type: ring, necklace, bracelet
Material: gold-plated sterling silver, gold, platinum, horse tail hair, cubic zirconia

器——双生，器——本源，器——容与放
WARE – TWIN, WARE – ORIGIN , WARE – CAPACITY AND RELEASE

作者姓名：许璐璐（中国）
作品类型：手镯，其他
作品材质：银，铜，树脂

Artist: Xu Lulu（China）
Type: bracelet, others
Material: silver, copper, resin

行云流水系列
FLOATING CLOUDS AND FLOWING WATER CATENA

作者姓名：闫政旭（中国）
作品类型：项饰，胸针
作品材质：纯银

Artist: Yan Zhengxu (China)
Type: necklace, brooch
Material: fine silver

乐家园，失乐园
HAPPY HOME , PARADISE LOST

作者姓名：杨井兰（中国）
作品类型：项饰
作品材质：银，白水晶

Artist: Yang Jinglan (China)
Type: necklace
Material: silver, white crystal

对话
DIALOGUE

作者姓名：杨漫（中国）
作品类型：项饰
作品材质：24k 金，白玉，铝

Artist: Yang Man（China）
Type: necklace
Material: 24ct gold, white jade, aluminu

丈量——手指的距离系列，
丈量——关系的距离系列，
探知身体空间——手指的舞伴

MEASUREMENT – FINGER DISTANCE SERIES，
MEASUREMENT – DISTANCE SERIES OF RELATIONSHIPS，
EXPLORE BODY SPACE – FINGER PARTNER

作者姓名：杨晓晖（中国）
作品类型：戒指
作品材质：银，纸，铝

Artist: Yang Yilun（China）
Type: ring
Material: gold, silver, paper, ename

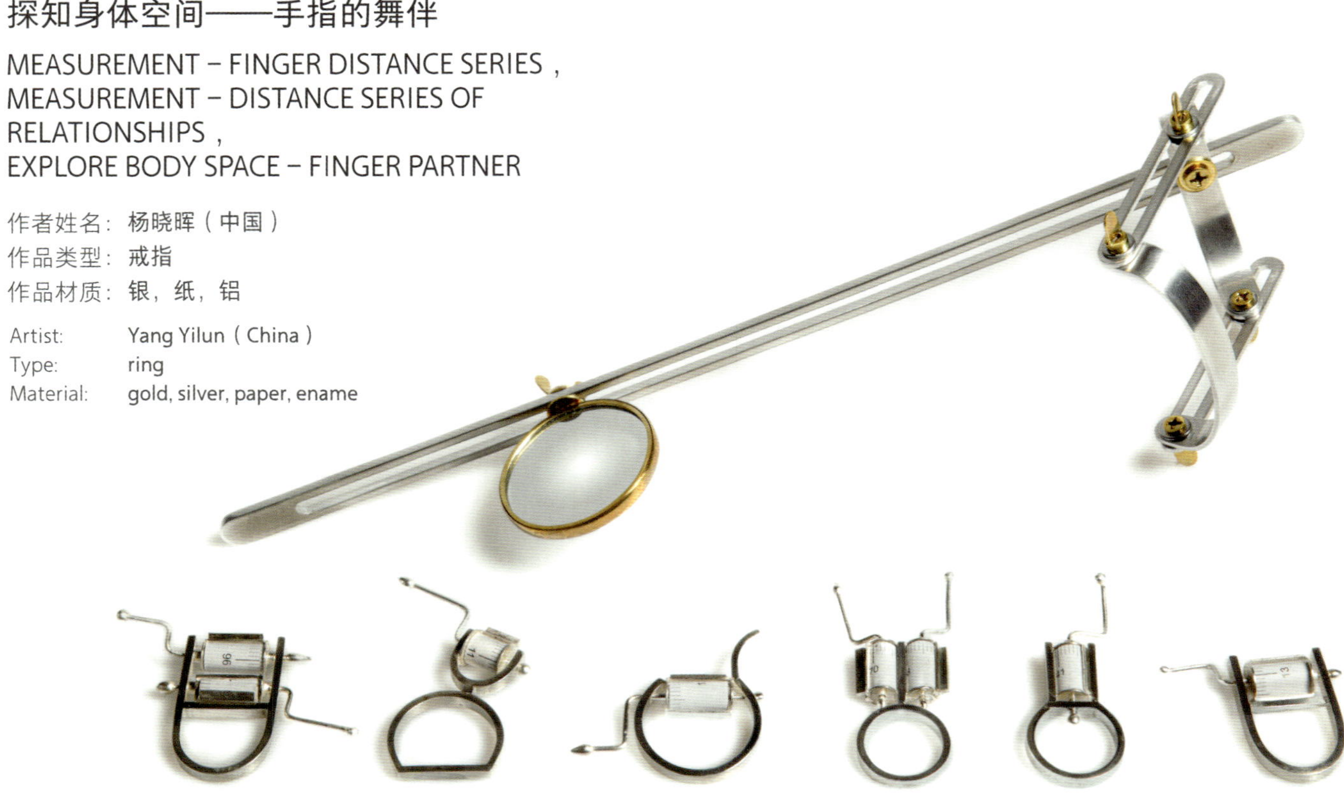

暗流——Ⅰ
UNDERCURRENT-Ⅰ

作者姓名：杨逸伦（中国）
作品类型：胸针
作品材质：金，银，珐琅

Artist: Yang Yilun（China）
Type: necklace
Material: gold, silver, enamel

盛开
BLOOMING

作者姓名：杨钊（中国）
作品类型：胸针
作品材质：银，黄铜，不锈钢

Artist: Yang Zhao（China）
Type: brooch
Material: silver, brass, stainless steel

欲
DESIRE

作者姓名：姚瑶（中国）
作品类型：项饰
作品材质：铜，宝石，石膏，纤

Artist: Yao Yao (China)
Type: necklace
Material: copper, gemstones, gypsum, fibers

光的沙漏 · 墨波
THE LIGHT OF THE HOURGLASS · THE INK WAVE

作者姓名：叶秀薇（中国）
作品类型：项饰
作品材质：铜线

Artist: Ye Xiuwei（China）
Type: necklace
Material: copper wire

黄色枕头，脸：冷漠，心跳
YELLOW PILLOW，FACE: INDIFFERENT，HEARTBEAT

作者姓名：尹蔡扬（中国）
作品类型：袖扣
作品材质：纯银，食品级硅胶，细银，不锈钢

Artist: Yin Caiyang（China）
Type: cufflinks
Material: fine silver, foodclass silicone rubber, fine silver, stainlesss steel

空想解剖——长臂金龟
ANATOMY OF A FANTASY LONG-ARMED TORTOISE

作者姓名：尹相锟（中国）
作品类型：胸针
作品材质：18k 金，海蓝宝石

Artist: Yin Xiangkun（China）
Type: brooch
Material: 18ct gold, aquamarine

“怡”系列
"YI" SERIES

作者姓名：于芳（中国）
作品类型：胸针，项饰
作品材质：银，矿物水晶，珍珠，黄花梨等

Artist: Yu Fang（China）
Type: brooch, necklace
Material: silver, mineral crystal, pearl, yellow pear wood, etc.

爆米花戒指
POPCORN RING

作者姓名：余诗颖（中国）
作品类型：戒指等
作品材质：银，轻黏土

Artist:	Yu Shiying (China)
Type:	ring, etc.
Material:	silver, light clay

午马
ZODIAC HORSE

作者姓名：袁春然（中国）
作品类型：胸针
作品材质：银，珐琅

Artist: Yuan Chunran（China）
Type: brooch
Material: sliver, enamel

看雾，雾色
EXPERIENCE FOG , COLOR FOG

作者姓名：翟悉涵（中国）
作品类型：头饰，项链
作品材质：塑料，银

Artist: Zhai Xihan (China)
Type: headwear, necklace
Material: plastic, silver

黑色花园
BLACK GARDEN

作者姓名：张帆（中国）
作品类型：戒指
作品材质：银

Artist: Zhang Fan（China）
Type: ring
Material: silver

旋玉蟠龙
XUAN YU PAN LONG

作品姓名：张凡，高松峰（中国）
作品类型：项链
作品材质：和田玉，紫铜鎏足金，足金

Artist: Zhang Fan , Gao Songfeng (China)
Type: necklace
Material: nephrite, gold-plated purple copper, pure gold

怀忆
MISS MEMORIES

作者姓名：张琨（中国）
作品类型：头饰
作品材质：俄罗斯软玉

Artist: Zhang Kun (China)
Type: headwear
Material: Russian nephrite

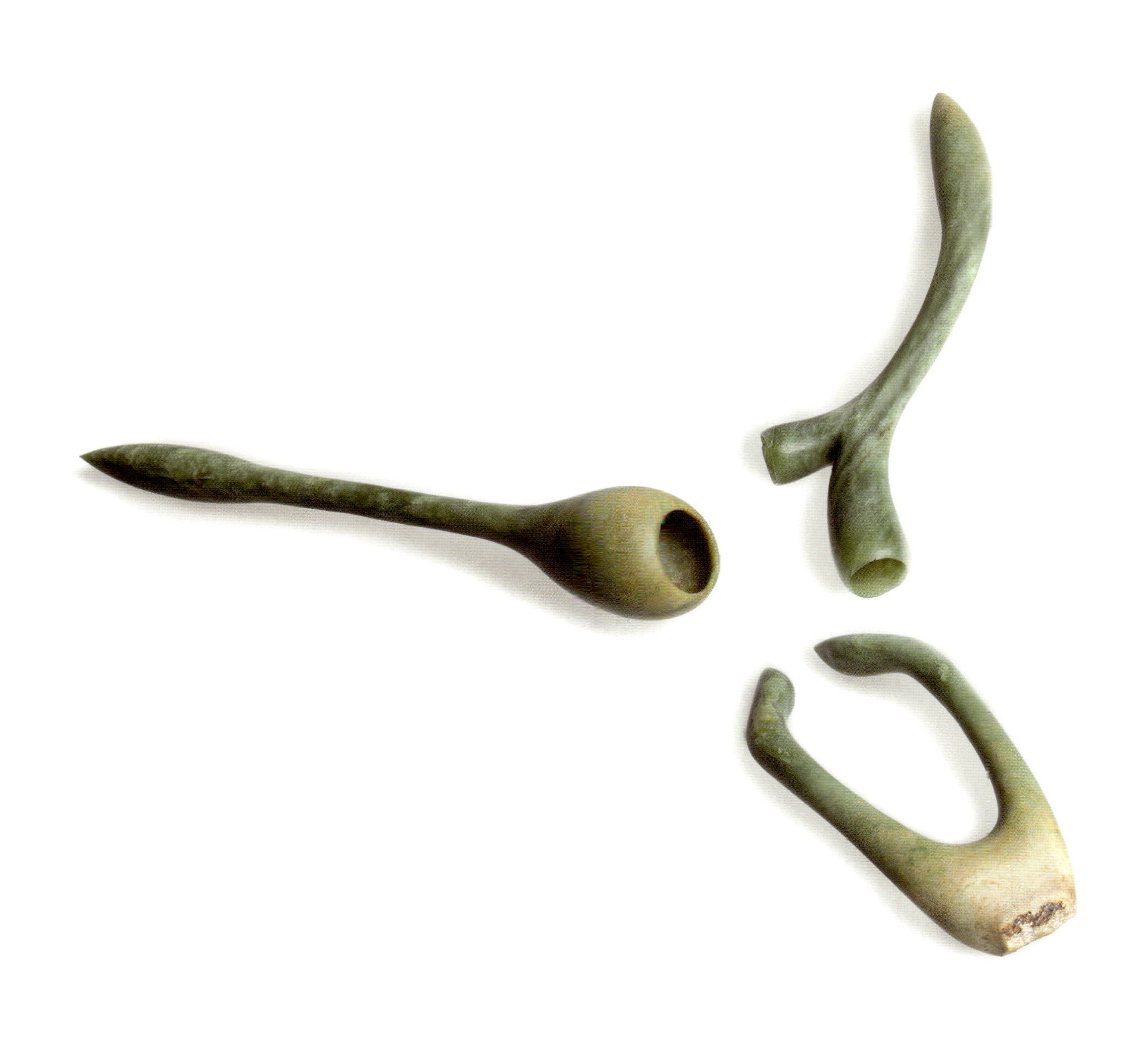

多面体
POLYHEDRON

作者姓名：张莉（中国）
作品类型：耳坠，胸针
作品材质：银，晶石

Artist: Zhang Li (China)
Type: earring, brooch
Material: silver, spar

亘古·永恒琥珀系列
ETERNAL AMBER SERIES

作者姓名：张荣红（中国）
作品类型：胸针吊坠两用
作品材质：加里宁琥珀，925 银

Artist: Zhang Ronghong (China)
Type: brooch pendant dual-purpose
Material: calinin amber, sterling silver

光明坛城系列套装，寂静系列套装
GUANGMING TANCHENG SERIES SETS, SILENCE SERIES SETS

作者姓名：张伟（中国）
作品类型：戒指，项饰，手镯，吊坠，耳饰
作品材质：18k 金，天然漆，蓝宝石，钻石

Artist: Zhang Wei（China）
Type: ring, necklace, bracelet, pendant, earrings
Material: 18ct gold , lacquer, sapphire, diamond

秋意
AUTUMN

作者姓名：张伟峰（中国）
作品类型：戒指，耳饰
作品材质：银，银镀金，氧化银，石榴石

Artist: Zhang Weifeng（China）
Type: ring, earring
Material: silver, gold-plated silver, oxide silver, garnet

三合一
THREE IN ONE

作者姓名：张雯（中国）
作品类型：摆件
作品材质：银

Artist: Zhang Wen （China）
Type: decoration
Material: silver

生命的旅程系列
THE JOURNEY OF LIFE SERIES

作者姓名：张雯迪（中国）
作品类型：胸针
作品材质：银，铜，珐琅，纤维

Artist: Zhang Wendi (China)
Type: brooch
Material: silver, copper, enamel, fiber

盔犀鸟之殇，鱼的世界，战后
THE WOUND OF THE HELMETED HORNBILL, THE WORLD OF THE FISH , AFTER THE WAR

作者姓名：张潇娟（中国）
作品类型：项饰
作品材质：黑檀，白银，玛瑙

Artist: Zhang Xiaojuan (China)
Type: necklace
Material: ebony, silver, agate

解忧发射器 I 号
SAD EJECTOR NO.1

作者姓名：张植（中国）
作品类型：胸针，其他
作品材质：银，铜，珐琅，纤维

Artist: Zhang Zhi（China）
Type: brooch, others
Material: silver, copper, enamel, fiber

我有一颗金牙
I HAVE A GOLD TOOTH

作者姓名：章藻藻（中国）
作品类型：胸针吊坠两用
作品材质：银，黄金，钻石

Artist: Zhang Zaozao（China）
Type: brooch pendant dual purpose
Material: silver, gold, diamond

微笑糖果
SMILE CANDY

作者姓名：赵慧颖（中国）
作品类型：胸针
作品材质：银，珐琅，亚克力

Artist: Zhao Huiying（China）
Type: brooch
Material: silver, enamel, acrylic

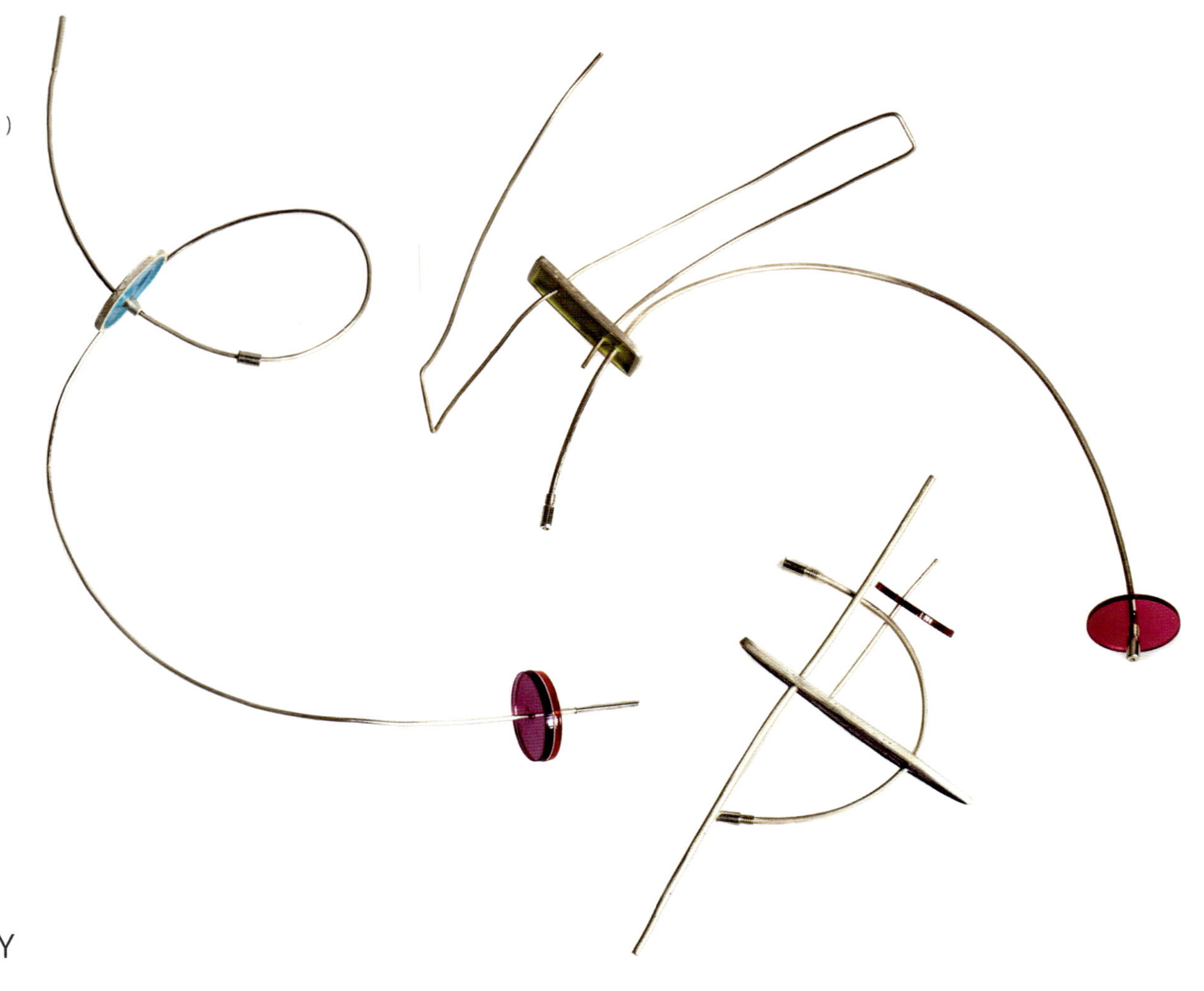

构城
CONSTRUCT THE CITY

作者姓名：赵剑侠（中国）
作品类型：戒指
作品材质：925 银，建筑废弃物

Artist: Zhao Jianxia（China）
Type: ring
Material: sterling silver, construction waste

燕子风筝（红色）
THE SWALLOW KITE（RED）

作者姓名：赵世笺（中国）
作品类型：胸针
作品材质：925 银，回收木材（染色）

Artist: Zhao Shijian（China）
Type: brooch
Material: sterling silver, recycled wood（dye）

光影
SHADOW

作者姓名：赵晓穆（中国）
作品类型：项饰
作品材质：黑玛瑙，砗磲，黄铜

Artist: Zhao Xiaomu（China）
Type: necklace
Material: black agate, giant clam, brass

交响
SYMPHONY

作者姓名：赵晔童（中国）
作品类型：项饰
作品材质：电路板，纽扣电池

Artist:	Zhao Yetong（China）
Type:	necklace
Material:	circuit board, button battery

澜色耳钉
LAN COLOR

作者姓名：赵祎（中国）
作品类型：耳饰
作品材质：乌木，天然漆，银

Artist: Zhao Yi（China）
Type: earring
Material: ebony, lacquer, silver

在血中奔跑，突变 8.1，眼睛 2.0
IT RUNS IN THE BLOOD , MUTATION 8.1 , EYE 2.0

作者姓名：赵英琪（中国）
作品类型：项链
作品材质：瓷器，石器，铜，925 银，釉瓷，电线

Artist: Zhao Yingi (China)
Type: necklace
Material: porcelain, stoneware, copper, sterling silver, vitreous enamel, wire

生净系列
SHENG JING SERIES

作者姓名：郑妍芳（中国）
作品类型：项饰
作品材质：银，竹，珍珠

Artist: Zheng Yanfang（China）
Type: necklace
Material: sliver, bamboo, pearl

博物盒之一
ONE OF THE NATURAL HISTORY BOXES

作者姓名：钟奕（中国）
作品类型：胸针
作品材质：925 银镀金，天然漆

Artist: Zhong Yi（China）
Type: brooch
Material: gold-plated sterling silver, lacquer

云鸢，彩云追月
KITE , COLORFUL CLOUDS CHASING THE MOON

作者姓名：朱欢（中国）
作品类型：胸针，项链
作品材质：银，珍珠，冷珐琅

Artist: Zhu Huan（China）
Type: brooch, necklace
Material: silver, pearl, cold enamel

镜
MIRROR

作者姓名：庄冬冬（中国）
作品类型：胸针
作品材质：镭射膜，亚克力，银

Artist: Zhuang Dongdong（China）
Type: brooch
Material: laser film, acrylic, silver

观察
OBSERVATION

作者姓名：邹艾耘（中国）
作品类型：胸针
作品材质：银，手绘热缩片

Artist: Zou Aiyun（China）
Type: brooch
Material: silver, hand-painted hot shrink film

头羊
HEAD SHEEP

作者姓名：邹宁馨（中国）
作品类型：胸针
作品材质：18k 金，黄铁矿，菊石，蓝宝石

Artist: Zou Ningxin（China）
Type: brooch
Material: 18ct gold, pyrite, ammonite, sapphire